T0134909

Studies in Systems, Decision and Control

Volume 100

Series editor

Janusz Kacprzyk, Polish Academy of Sciences, Warsaw, Poland
e-mail: kacprzyk@ibspan.waw.pl

About this Series

The series "Studies in Systems, Decision and Control" (SSDC) covers both new developments and advances, as well as the state of the art, in the various areas of broadly perceived systems, decision making and control- quickly, up to date and with a high quality. The intent is to cover the theory, applications, and perspectives on the state of the art and future developments relevant to systems, decision making, control, complex processes and related areas, as embedded in the fields of engineering, computer science, physics, economics, social and life sciences, as well as the paradigms and methodologies behind them. The series contains monographs, textbooks, lecture notes and edited volumes in systems, decision making and control spanning the areas of Cyber-Physical Systems, Autonomous Systems, Sensor Networks, Control Systems, Energy Systems, Automotive Systems, Biological Systems, Vehicular Networking and Connected Vehicles, Aerospace Systems, Automation, Manufacturing, Smart Grids, Nonlinear Systems, Power Systems, Robotics, Social Systems, Economic Systems and other. Of particular value to both the contributors and the readership are the short publication timeframe and the world-wide distribution and exposure which enable both a wide and rapid dissemination of research output.

More information about this series at http://www.springer.com/series/13304

Martine Ceberio · Vladik Kreinovich
Editors

Constraint Programming and Decision Making: Theory and Applications

 Springer

Editors
Martine Ceberio
Department of Computer Science
University of Texas at El Paso
El Paso, TX
USA

Vladik Kreinovich
Department of Computer Science
University of Texas at El Paso
El Paso, TX
USA

ISSN 2198-4182 ISSN 2198-4190 (electronic)
Studies in Systems, Decision and Control
ISBN 978-3-319-87153-0 ISBN 978-3-319-61753-4 (eBook)
DOI 10.1007/978-3-319-61753-4

Printed on acid-free paper

This Springer imprint is published by Springer Nature
The registered company is Springer International Publishing AG
The registered company address is: Gewerbestrasse 11, 6330 Cham, Switzerland

Preface

Constraint programming and decision making are important. Constraint programming and decision making techniques are essential in the building of intelligent systems. They constitute an efficient approach to representing and solving many practical problems. They have been applied successfully to a number of fields, such as scheduling of air traffic, software engineering, networks security, chemistry, and biology. However, despite the proved usefulness of these techniques, they are still underutilized in real-life applications. One reason is the perceived lack of effective communication between constraint programming experts and domain practitioners about constraints, in general, and their use in decision making, in particular.

CoProd workshops. To bridge this gap, annual International Constraint Programming and Decision Making workshops CoProd'XX have been organized since 2008: in El Paso, Texas (2008, 2009, 2011, 2013, and 2015), in Lyon, France (2010), in Novosibirsk, Russia (2012), and in Würzburg, Germany (2014); CoProd'2016 was held in Uppsala, Sweden. Papers from the previous workshops appeared in [8]. This volume contains extended version of selected papers presented at the following CoProd workshops.

CoProD workshops aim to bring together, from areas closely related to decision making, researchers who design solutions to decision making problems and researchers who need these solutions and likely already use some solutions. Both communities are often not connected enough to allow cross-fertilization of ideas and practical applications.

CoProD workshops aim at facilitating networking opportunities and cross-fertilization of ideas between the approaches used in the different attending communities. Because of this, in addition to active researchers in decision making and constraint programming techniques, these workshops are also attended by domain scientists—whose participation and input is highly valued in these workshops.

The *goal of CoProD workshops is therefore to constitute a forum for inter-community building*. The objectives of this forum are to facilitate:

- The presentation of advances in constraint solving, optimization, decision making, and related topics;
- The development of a network of researchers interested in constraint techniques, in particular researchers and practitioners that use numeric and symbolic approaches (or a combination of them) to solve constraint and optimization problems;
- The gap bridging between the great capacity of the latest decision making/constraint techniques and their limited use.

CoProD workshops can impact these communities by easing collaborations and therefore the emergence of new techniques, and by creating a network of interest. The objectives of CoProD are also relayed all year round through the Web site constraintsolving.com.

Topics of interest. The main emphasis is on the joint application of constraint programming and decision making techniques to real-life problems. Other topics of interest include:

- Algorithms and applications of:

  - Constraint solving, including symbolic-numeric algorithms
  - Optimization, especially optimization under constraints (including multi-objective optimization)
  - Interval techniques in optimization and their interrelation with constraint techniques
  - Soft constraints
  - Decision making techniques

- Description of domain applications that:

  - Require new decision making and/or constraint techniques
  - Implement decision making and/or constraint techniques

Contents of the present volume: general overview. All these topics are represented in the papers forming the current volume. These papers cover all the stages of decision making under constraints:

- How to formulate the problem of decision making in precise terms, taking different criteria into account?
- How to check whether (and when) the corresponding decision problem is algorithmically solvable?
- Once we know that the decision problem is, in principle, algorithmically solvable, how to find the corresponding algorithm, and how to make this algorithm as efficient as possible?
- How to take into account uncertainty, whether it is given in terms of bounds (intervals), probabilities, or fuzzy sets?

How to formulate the problem of decision making in precise terms: case of single-agent multi-criterion decision making. Paper [3] shows that in many cases, we can efficiently formulate single-agent multi-criterion decision making problems in terms of constraints, and thus, known constraint-based techniques to solve these problems.

How to formulate the problem of decision making in precise terms: general case. In the general case, in addition to several criteria, we may also have several agents. There are two important aspects of multi-agent problems:

- First, even when different agents have similar interests, their estimates of the values of different criteria are often drastically different; the papers [2, 11] analyze how to best reconcile these differences and come up with a reasonable solution;
- Second, different agents may have different interests; such situations are analyzed in [14].

When are problems algorithmically solvable? It is known that the corresponding problems stop being solvable if there is a discontinuity. Interestingly, it turns out that for many problems, discontinuity is the only obstacle to algorithmic solvability; see, e.g., [6]. Even in the discontinuous case, many problems are algorithmically solvable in some weaker—but still physically meaningful—sense [7].

While some of these problems are algorithmically solvable, many of them are NP-hard, meaning that—unless P=NP—no feasible algorithm is possible that would always compute the exact solution to the corresponding problem; see, e.g., [5].

How to design efficient algorithms for solving the problems. In some cases, there already are algorithms for solving similar problems, but these algorithms only work under difficult-to-test assumptions. For example, many efficient algorithms rely on global assumptions about the problems, assumptions which—due to their global character—are difficult to check. It has been empirically determined that in many such cases, there is a version that only depends only on local (thus, easier-to-test) constraints. The paper [4] provides a general theoretical explanation of this result and a general algorithm transforming global constraint results into the corresponding local constraint ones.

In other cases, we do not have ready algorithms. In such cases, it is reasonable to see how we humans solve problems, and to borrow the corresponding ideas. This can be done on several levels: It can be done on the higher level, by simulating how we reason, or at a deeper biological level, by simulating how the brain works when we solve such problems.

On the reasoning level, one of the most efficient ways of how we humans solve problem is that we ignore unnecessary details and thus go to a certain level of abstraction. This is a trade-off: If we ignore too many important details, the solution becomes too far from optimal, but if we leave too many unnecessary details, the resulting requires too many computations; there needs to be an optimal level of abstraction. There is an empirical approach to finding such level, called similarity approach. The paper [16] provides a theoretical explanation for this approach.

On the biological level, it has indeed turned out to be computationally efficient to use neural networks, i.e., algorithms that emulate how our brain's neurons work. The paper [1] provides a theoretical explanation for this empirical success.

How to take uncertainty into account. The simplest type of uncertainty is an interval uncertainty, when instead of the exact value of a quantity x we only know its lower bounds $\underline{x}$ and its upper bound $\bar{x}$—i.e., the interval $[\underline{x}, \bar{x}]$ that contains this value.

Interestingly, interval uncertainty is in good accordance with how we often evaluate our experience: by only taking into account the largest possible value $\bar{x}$ and the last occurred value; the reasons behind such an evaluation are shown in [15]. Similarly, it turns that a good strategy in predicting the behavior of stock markets is to ignore its fluctuation and to only take into account its (local) minima and maxima; a theoretical explanation for such a strategy is given in [18].

In some situations, we only know the bounds on the quantities. In other situations, we know how these quantities depend on certain parameters—but we know the values of these parameters only with interval uncertainty. In particular, in control situations, the dynamics of a system is often described by a matrix whose dependence on several parameters is known, but for which the values of these parameters are only known with interval uncertainty [12] provides efficient algorithms for checking whether an important property like positive definiteness holds for all possible values if the corresponding parameters.

In addition to the interval of possible values, we may know the probabilities of different values within this interval (probabilistic approach) or, if we do not know these probabilities, the expert evaluations of how possible these values are (fuzzy approach).

In case of the probabilistic approach, one of the main problems is determining these probabilities. For the case of measurements, this problem is analyzed in [17].

For the case of fuzzy uncertainty, constraint optimization problems are analyzed in [10].

Known techniques of solving the corresponding problems are practically useful, but these techniques often involve making rather arbitrary choices that affect the result. For example, a known method of optimization under fuzzy constraints—proposed originally by L. Zadeh, the father of fuzzy logic, and by R. Bellman, a renowned specialist in optimization and control—strongly depends on the rather arbitrary selection of the unconstrained maximum. The paper [9] analyzed when the resulting solution does not depend on this selection.

Resulting applications. Papers presented in this volume include numerous applications, including applications:

- To control [12, 13]: how to take into account interval uncertainty,
- To economics [18]: how to predict stock market behavior,
- To environmental sciences and geosciences [3]: how to combine data of different types,
- To manufacturing [14]: how to optimally determine the production level.

Thanks. We are greatly thankful to National Science Foundation for supporting several CoProd workshops, to all the authors and referees, and to all the participants of the CoProd workshops. Our special thanks to Prof. Janusz Kacprzyk, the editor of this book series, for his support and help. Thanks to all of you!

El Paso, USA   Martine Ceberio
September 2017   Vladik Kreinovich

References

1. Baral, C., Fuentes, O., Kreinovich, V.: Why deep neural networks: a possible theoretical explanation (this volume)
2. Bistarelli, S., Ceberio, M., Henderson, J.A., Santini, F.: Abstract argumentation frameworks to promote fairness and rationality in multi-experts multi-criteria decision making (this volume)
3. Ceberio, M., Kosheleva, O., Kreinovich, V.: Constraint approach to multi-objective optimization (this volume)
4. Ceberio, M., Kosheleva, O., Kreinovich, V.: From global to local constraints: a constructive version of bloch's principle (this volume)
5. Ceberio, M., Kosheleva, O., Kreinovich, V.: Optimizing pred(25) is NP-hard (this volume)
6. Ceberio, M., Kosheleva, O., Kreinovich, V.: Range estimation under constraints is computable unless there is a discontinuity (this volume)
7. Ceberio, M., Kosheleva, O., Kreinovich, V.: Towards a physically meaningful definition of computable discontinuous and multi-valued functions (constraints) (this volume)
8. Ceberio, M., Kreinovich, V. (eds.): Constraint Programming and Decision Making. Springer, Berlin, Heidelberg (2014)
9. Figueroa Garcia, J.C., Ceberio, M., Kreinovich, V.: Algebraic product is the only t-norm for which optimization under fuzzy constraints is scale-invariant (this volume)
10. Figueroa Garcia, J.C., Hernandez-Perez, G., Kalenatic, D.: Comparing operation points in linear programming with fuzzy constraints (this volume)
11. Garbayo, L., Ceberio, M., Bistarelli, S., Henderson, J.: On modeling multi-experts multi-criteria decision-making argumentation and disagreement: philosophical and computational approaches reconsidered (this volume)
12. Hladik, M.: Positive semidefiniteness and positive definiteness of a linear parametric interval matrix (this volume)
13. Jeyasenthil, R., Nataraj, P.S.V., Purohit, H.: Automatic loop-shaping of H_∞/μ problem in QFT using interval consistency based hybrid optimization (this volume)
14. Kalashnikov, V.V., Bulavsky, V.A., Kalashnykova, N.I.: Existence of Nash-optimal strategies in themeta-game (this volume)
15. Kosheleva, O., Ceberio, M., Kreinovich, V.: Peak-end rule: a utility-based explanation (this volume)
16. Lorkowski, J., Trnecka, M.: Similarity approach to defining basic level of concepts explained from the utility viewpoint (this volume)
17. Servin, C., Kreinovich, V.: Comparisons of measurement results as constraints on accuracies of measuring instruments: when can we determine the accuracies from these constraints? (this volume)
18. Stylios, C.D., Kreinovich, V.: Dow theory's peak-and-trough analysis justified (this volume)

Contents

Why Deep Neural Networks: A Possible Theoretical Explanation

Chitta Baral, Olac Fuentes and Vladik Kreinovich

1 Formulation of the Problem

Why neural networks. In spite of all the progress in computer-based recognition algorithms, we humans still perform many recognition tasks much faster (and often much more reliably) than computer programs. And we perform faster in spite of the fact that the fastest of our brain's data processing units—neurons—has reaction time ≈ 10 msec, while computer components operate in nanoseconds. The explanation lies largely in the fact that in the human brain, billion of neurons operate in parallel.

Thus, a natural idea is to speed up computer-based data processing, by simulating the way biological neurons operate. The resulting data processing techniques are known as *artificial neural networks*, or simply *neural networks*, for short; see, e.g., [1].

Traditionally, neural networks used the smallest possible number of layers. When we have neurons working in parallel, the computation time is proportional to the number of layers that the signal passes through:

- in each layer, all the processing is done in parallel,
- so, data processing in each layer takes the same time, no matter how many neurons we use.

C. Baral (✉)
Department of Computer Science and Engineering, Arizona State University,
Tempe, AZ 85287-5406, USA
e-mail: baral@asu.edu

O. Fuentes · V. Kreinovich
Department of Computer Science, University of Texas at El Paso, 500 W. University,
El Paso, TX 79968, USA
e-mail: ofuentes@utep.edu

V. Kreinovich
e-mail: vladik@utep.edu

© Springer International Publishing AG 2018
M. Ceberio and V. Kreinovich (eds.), *Constraint Programming and Decision Making: Theory and Applications*, Studies in Systems, Decision and Control 100,
DOI 10.1007/978-3-319-61753-4_1

Most widely used neurons perform two types of operations: a linear combination of inputs $y = w_0 + \sum_{i=1}^{n} w_i \cdot x_i$ and a non-linear transformation $y = s(x)$ for some non-linear *activation function* $s(x)$. Activations functions are usually assumed to be smooth (at least three times differentiable). The most widely used activation function is the sigmoid function $s(x) = \dfrac{1}{1 + \exp(-x)}$.

It is known that, for the sigmoid activation function, already 3-layer neurons are universal approximators; see, e.g., [1]. To be more precise, the following class of functions can approximate any continuous function $f(x_1, \ldots, x_n)$ on a given box $[\underline{x}_1, \overline{x}_1] \times \ldots \times [\underline{x}_n, \overline{x}_n]$ with a given accuracy ε:

$$y = \sum_{k=1}^{K} W_k \cdot y_k - W_0, \tag{1}$$

where

$$y_k = s(z_k) \tag{2}$$

and

$$z_k = \sum_{i=1}^{n} w_{ki} \cdot x_i - w_{k0}. \tag{3}$$

In such neurons:

- the original signals x_i pass through the first linear layer in which all the values z_k are computed;
- then the second (non-linear) layer computes all the values y_k, and
- finally, the third (linear) layer computes the resulting value y.

It is also known that 2-layer neural networks do not have the universal approximation property. As a result, 3-layer networks used to be most frequently used.

Recent successes of deep networks: a mystery. Recently, it was empirically shown that in many cases, it is beneficial to use "deep" neural networks, i.e., neural networks with a large number of layers; see, e.g., [2, 3, 5–7]. What is still not clear is why this works better than the more traditional (and seemingly better) 3-layer network.

Comment To be more precise, there are qualitative explanations for this empirical phenomena, but they have not been transformed into a precise result:

- One qualitative explanation is that if we have few neurons on each layer, then we have fewer combinations of weights on each layer, so it is easier to try all such combinations.
- Another qualitative explanation is that when we have several neurons on the same layer, there is a potential duplication of information—since we can have two identical neurons—but neurons on different layers do not lead to duplication.

2 Why Deep Neural Networks: Our Explanation

The universal approximation property of 3-layer networks depends on the choice of the activation function: reminder. In principle, different activation functions $s(x)$ are used in neural networks. The most important requirement is that the function $s(x)$ should be non-linear: otherwise, we will only be able to represent linear functions.

The universal approximation result for 3-layer networks was originally proved for the sigmoid activation function. A similar result is true for many other activations functions, but it is not true for many other non-linear functions. For example, if we use a non-linear function $s(x) = x^2$, then the network (1)–(3) is only able to compute quadratic functions—and thus, it will not have the universal approximation property.

Similar results. Similarly, if we select $s(x)$ to be any polynomial $s(x) = a_0 \cdot x^d + a_1 \cdot x^{d-1} + \ldots + a_{d-1} \cdot x + a_d$, then every function computed by a network (1)–(3) is a polynomial of degree $\leq d$, and thus, the corresponding network does not have the universal approximation property either.

A similar negative result holds even if, instead of a 3-layer network, we allow multi-layer networks with the possibility of ℓ non-linear layers. Indeed:

- the function computed by the network is a composition of functions corresponding to each layer; and
- the composition $P_1(P_2(x))$ of two polynomials $P_1(x)$ and $P_2(x)$ of degrees d_1 and d_2 has a degree $d_1 \cdot d_2$.

Thus, if the activation function is a polynomial of degree d, and we allow ℓ non-linear layers, then each function computed such a network is a polynomial of degree $\leq D \stackrel{\text{def}}{=} d^\ell$. Thus, such networks are not universal approximators.

Why this is important. Since we are mostly using the sigmoid activation function $s(x)$, why does it matter that something is wrong with other functions $s(x)$? At first glance, the above negative results only emphasize the importance of using the sigmoid activation function.

In the ideal world, yes. However, in reality, no matter how we implement the activation function, whether we implement it in hardware or in software, we cannot implement is exactly. We can implement the activation function with a certain accuracy. As a result, in a real neural network, instead of the desired sigmoid activation function $s(x)$, we actually have an approximate function $\widetilde{s}(x) \approx s(x)$.

And here lies a problem. It is known (see, e.g., [1]) that an arbitrary continuous function on a box can be approximated, within any given accuracy, by a polynomial. This is not just a theoretical possibility: when a computer computes a standard non-linear function, be it $\sin(x)$ or $\exp(x)$, the most widely use computational algorithms actually compute the sum of the first few terms in the Taylor expansion of the desired function—i.e., actually compute a polynomial, the activation function for which the universal approximation property is lost.

Conclusion: the usual formulation of the universal approximation property is not fully adequate. The usual formulation of the universal approximation property assumes that we can implement the exact activation function. In practice, however, we can only implement some approximation to the ideal activation function—and, if we take that into account, that the universal approximation property may be lost.

To study real-life neural networks, it is therefore desirable to come up with a more adequate formulation, that takes into account the fact that an activation function can only be implemented with a certain accuracy. Let us formulate this in precise terms.

Definition 1 By a ℓ-layer neural network with activation function $s(x)$ and n inputs, we mean a (marked) ordered graph whose vertices (called *neurons*) are divided into $\ell + 1$ subsets (called *layers*) 0, 1, ..., ℓ, in such a way that:

- the 0-th (*input*) layer has exactly n neurons marked $x_1, \ldots, x_n$;
- the last (ℓ-th) layer has exactly one neuron marked y;
- an edge from a neuron in the i-the layer can only go to a neuron in a j-th layer, with $j > i$;
- some neurons who have only one incoming edge are marked by s; each such neuron applies the activation function $s(z)$ to the output z of the incoming neuron;
- for each neuron that is not marked by s, each edge going into this neuron is marked by a real number w_i and the neuron itself is marked by a number w_0; this neuron computes the value $\sum_i w_i \cdot y_i - w_0$, where y_i are the outputs of the incoming neurons.

The markings enable us to compute, layer-by-layer, from Layer 1 to Layer ℓ, the output of each neuron, until we reach the output of the neuron in the final layer; its output is called the *result of applying the neural network to the inputs* $x_1, \ldots, x_n$.

Definition 2 Let $\delta > 0$ be a real number. We say that functions $s(x)$ and $\widetilde{s}(x)$ defined on an interval $[-X, X]$ are δ-*close* if $|\widetilde{s}(x) - s(x)| \leq \delta$ for all $x \in [-X, X]$.

Definition 3 Let $s(x)$ be a given smooth activation function. We say that a class of neural networks with this activation function has the *realistic universal approximation property* if:

- for every continuous functions $f(x_1, \ldots, x_n)$ on a box $[\underline{x}_1, \overline{x}_1] \times \ldots \times [\underline{x}_n, \overline{x}_n]$,
- for every two real numbers $\delta > 0$ and $\varepsilon > 0$, and
- for every smooth function $\widetilde{s}(x)$ which is δ-close to $s(x)$,

there exists a neural network from this class for which,

- when we use the activation function $\widetilde{s}(x)$,
- for all inputs from the given box, the result of applying this neural network is ε-close to $f(x_1, \ldots, x_n)$.

Proposition 1 *For any ℓ and for any activation function $s(x)$, the class of all ℓ-layer neural networks does not have the realistic universal approximation property.*

Proof follows directly from the fact that we can approximate any function $s(x)$ by polynomials $\tilde{s}(x)$, and, and we have shown, ℓ-layer neural networks that use a polynomial activation function do not have the universal approximation property.

Proposition 2 *For any nonlinear activation function $s(x)$, the class of all neural networks has the realistic universal approximation property.*

Comment The proof of this result is, in effect, contained in [4], where it is shown that if we do not limit the number of layers, then any non-linear activation function has the universal approximation property.

3 Conclusion

In the idealized case, when we assume that we can implement the activation function exactly, 3-layer networks have the universal approximation property. However, in a more realistic setting, if we take into account that we can only implement the activation function approximately, neither 3-layer networks not network with any fixed number of layer have the corresponding realistic universal approximation property. Thus, to adequately approximate different dependencies, we have to consider networks with many layers—i.e., deep networks. Thus, this theoretical result explains the need for deep neural networks.

References

1. Bishop, C.M.: Pattern Recognition and Machine Learning. Springer, New York (2006)
2. Erhan, D., Bengio, Y., Courville, A., Manzagol, P.-A., Vincent, P., Bengio, S.: Why does unsupervised pre-training help deep learning? J. Mach. Learn. Res. **11**, 625–660 (2010)
3. Hinton, G.E., Osindero, S., Teh, Y.-W.: A Fast Learning Algorithm for Deep Belief Nets. Neural Comput. **18**, 1527–1554 (2006)
4. Kreinovich, V.: Arbitrary nonlinearity is sufficient to represent all functions by neural networks: a theorem. Neural Netw. **4**, 381–383 (1991)
5. Krizhevsky, A., Sutskever, I., Hinton, G. E.: ImageNet classification with deep convolutional neural networks. Adv. Neural Inf. Process. Syst. pp. 1097–1105 (2012)
6. Mnih, V., Kavukcuoglu, K., Silver, D., Rusu, A.A., Veness, J., Bellemare, M.G., Graves, A., Riedmiller, M., Fidjeland, A.K., Ostrovski, G., et al.: Human-level control through deep reinforcement learning. Nature **518**(7540), 529–533 (2015)
7. Ngiam, J., Khosla, A., Kim, M., Lee, H., Ng, A. Y.: Multimodal deep learning. In: Proceedings of the 28-th International Conference on Machine Learning ICML'2011, Bellevue, Washington, USA, June 28–July 2, (2011), pp. 265–272

Abstract Argumentation Frameworks to Promote Fairness and Rationality in Multi-experts Multi-criteria Decision Making

Stefano Bistarelli, Martine Ceberio, Joel A. Henderson
and Francesco Santini

1 Introduction

Expert analysis and decisions arguably provide high-quality and highly-valued support for action and policy making in a wide variety of fields, from social services, to medicine, to engineering, to grant funding committees, and so on. However, the use of experts can be prohibitive due to either lack of availability, high cost, or limited time frame for action—this is the case particularly more so in impoverished areas. As such, it is desirable to be able to replicate/predict such decisions when beneficial even in the absence of experts. Unfortunately, there are many obstacles that still hinder an accurate simulation of expert decisions. First, it is hard to understand, and therefore replicate, the way each expert "aggregates"information/assessment along several criteria. In addition, even if we had a reasonable insight about it, any expert may make inconsistent decisions across similar scenarios. Finally, in the case of multiple experts, despite looking at the same information, two (or more) experts may disagree on the decisions to be made.

S. Bistarelli (✉)
Dipartimento di Matematica e Informatica, Università di Perugia,
Via Vanvitelli, 06123 Perugia, Italy
e-mail: bista@dmi.unipg.it

M. Ceberio · J.A. Henderson
Computer Science Department, The University of Texas at El Paso,
500 West University, El Paso, TX 79968, USA
e-mail: mceberio@utep.edu

J.A. Henderson
e-mail: jahenderson@miners.utep.edu

F. Santini
Istituto di Informatica e Telematica, CNR-Pisa, Via Moruzzi, 56124 Pisa, Italy
e-mail: francesco.santini@iit.cnr.it

© Springer International Publishing AG 2018
M. Ceberio and V. Kreinovich (eds.), *Constraint Programming and Decision
Making: Theory and Applications*, Studies in Systems, Decision and Control 100,
DOI 10.1007/978-3-319-61753-4_2

In spite of such challenges, traditional approaches seek to combine prior known decisions of experts into a classification of scenarios (machine learning approaches) or into some aggregation function that allows to best replicate the experts' decisions. Unfortunately, this line of approaches tends to overlook the irrationality and/or lack of fairness of experts, aggregating all available prior information regardless of quality.

In this work, we propose to model Multi-Experts Multi-Criteria Decision-Making (MEMCDM) problems using argumentation frameworks. We specifically design our proposed model so as to emulate fairness and rationality in decisions. For instance, when, of two expert's decisions, one is unfair, we impose an attack between these two decisions, forcing one of the two decisions out of the argumentation network's resulting extensions. Similarly, we specifically put irrational decisions in opposition to force one out. In doing so, we aim to enable the prediction of decisions that are themselves fair and rational. Our model is illustrated on two toy examples.

In what follows, we start by recalling preliminary notions, then we proceed with describing our model in details and illustrate our model in the case of Software Quality Assessment by multiple experts along multiple criteria.

2 Preliminary Notions

2.1 Multi-criteria Decision Making (MCDM)

Multi-criteria decision-making (MCDM) involves selecting one of several different alternatives, based on a set of criteria that describe the alternatives. However, there are numerous problems that make comparing these alternatives difficult. For instance, very often, decisions are based on several conflicting criteria; e.g., which car to buy that is cheap and energy efficient. In addition, what happens when we have a group of decision makers that must come to some sort of consensus? This is known as multi-expert multi-criteria decision making (MEMCDM). In MEMCDM, there are several new problems to be addressed. One such problem is how to handle expert disagreement and come to a consensus/decision in the first place. Another problem, as stated earlier, is that of predicting future decisions based on decision data from multiple experts along multiple criteria. Again, the question of "which expert/decision-making process to follow?" is a major challenge in solving such problems.

2.1.1 Approaches to MCDM

In general, on a daily basis, when the decision is not critical, in order to reach a decision, we mentally "average/sort"these criteria along with their satisfaction levels. This corresponds to aggregating values of satisfaction with weights on each criterion, reflecting its importance in the overall score (a.k.a. additive aggregation), that is, calculating the overall score of an alternative with the weighted sum of the criterion scores. In other words, weights assigned to different sets of criteria in the weighted-average approach form an "additive measure". Additive aggregation, however, assumes that criteria are independent, which is seldom the case [5]. Non-linear approaches also prove to lead to solutions that are not completely relevant [9].

This should change when considering possible dependence between criteria. For example, if two criteria are strongly dependent, it means that both criteria express, in effect, the same attribute. As a result, when we consider the set consisting of these two criteria, we should assign to this set the same weight as to each of these criteria—and not double the weight as in the weighted sum approach. In general, the weight associated to different sets should be different from the sum of the weights associated to individual criteria. In mathematics, such non-additive functions assigning numbers to sets are known as non-additive (fuzzy) measures. It is therefore reasonable to describe the dependence between different criteria by using an appropriate non-additive (fuzzy) measure. Combining the fuzzy measure values with the criteria satisfaction can be done using the Choquet integral, which integrals are actively used in Multi-Criteria Decision Making [8].

However, to make this happen, fuzzy measures need to be determined: they can either be identified by a decision maker/expert or by an automated system that extracts them from sample data. Since human expertise might not always be available and getting accurate fuzzy values (even from an expert) might be tedious [14], fuzzy measures are usually automatically extracted from prior decision decision data. To the original problem, this approach adds an optimization problem that can be tedious to solve. Although it was solved with success for some data sets [13], the overall prediction quality is not satisfactory and the approach limits the number of criteria that can be taken into account (the number of variables to determine is exponential in the number of criteria) [12].

2.2 Argumentation Frameworks

In this section we briefly summarise the background information related to classical AAFs [7]. We focus on the basic definition of an AAF (see Definition 1), on the notion of defence (Definition 2), and on extension-based semantics (Definition 3).

Definition 1 *(Abstract Argumentation Frameworks)* An Abstract Argumentation Framework (AAF) is a pair $F = \langle A, R \rangle$ of a set A of arguments and a binary relation $R \subseteq A \times A$, called the attack relation. $\forall a, b \in A$, $a R b$ (or, $a \rightarrowtail b$) means that a attacks b. An AAF may be represented by a directed graph (an interaction graph) whose nodes are arguments and edges represent the attack relation. A set of arguments $S \subseteq A$ attacks an argument a, i.e., $S \rightarrowtail a$, if a is attacked by an argument of S, i.e., $\exists b \in S.b \rightarrowtail a$.

Definition 2 *(Defence)* Given an AAF, $F = \langle A, R \rangle$, an argument $a \in A$ is defended (in F) by a set $S \subseteq A$ if for each $b \in A$, such that $b \rightarrowtail a$, also $S \rightarrowtail b$ holds. Moreover, for $S \subseteq A$, we denote by S_R^+ the set $S \cup \{b \mid S \rightarrowtail b\}$.

The "acceptability" of an argument [7] depends on its membership to some sets, called *extensions*: such extensions need to satisfy the properties required by a given semantics, and they characterise a collective "acceptability". In the following, *stb*, *adm*, *prf*, *gde*, *com*, and *sem*, respectively stand for stable, admissible, preferred, grounded, complete, and semi-stable semantics. The intuition behind these semantics is outside the scope of this work (e.g., see [10, Chap. 3]).

Definition 3 *(Semantics)* Let $F = \langle A, R \rangle$ be an AAF. A set $S \subseteq A$ is conflict-free (in F), denoted $S \in cf.(F)$, iff there are no $a, b \in S$, such that $(a, b), (b, a) \in R$. For $S \in cf.(F)$, it holds that:

- $S \in stb(F)$, if for each $a \in A \backslash S$, $S \rightarrowtail a$, i.e., $S_R^+ = A$;
- $S \in adm(F)$, if each $a \in S$ is defended by S;
- $S \in prf(F)$, if $S \in adm(F)$ and there is no $T \in adm(F)$ with $S \subset T$;
- $S = gde(F)$ if $S \in com(F)$ and there is no $T \in com(F)$ with $T \subset S$;
- $S \in com(F)$, if $S \in adm(F)$ and for each $a \in A$ defended by S, $a \in S$ holds;
- $S \in sem(F)$, if $S \in adm(F)$ and there is no $T \in adm(F)$ with $S_R^+ \subset T_R^+$.

We recall that for each AF, $stb(F) \subseteq sem(F) \subseteq prf(F) \subseteq com(F) \subseteq adm(F)$ holds, and that for each of the considered semantics σ (except stable) $\sigma(F) \neq \emptyset$ always holds. Moreover, in case an AF has at least one stable extension, its stable, and semi-stable extensions coincide. Finally, $gde(F)$ is always unique, and $gde(F) \in com(F)$.

Consider $F = \langle A, R \rangle$ in Fig. 1, with $A = \{a, b, c, d, e\}$ and $R = \{(a, b), (c, b), (c, d), (d, c), (d, e), (e, e)\}$. We have that $stb(F) = sem(F) = \{\{a, d\}\}$, and $gde(F) = \{a\}$. The admissible sets of F are $\emptyset, \{a\}, \{c\}, \{d\}, \{a, c\}, \{a, d\}$, and $prf(F) = \{\{a, c\}, \{a, d\}\}$. The complete extensions are $\{a\}, \{a, c\}, \{a, d\}$.

In the proposed model (precisely in Sect. 3.2) we take advantage of *symmetric AAFs* [6]:

Definition 4 *(Symmetric AAFs [6])* A symmetric (Abstract) Argumentation Framework is a finite Argumentation Framework $F = \langle A, R \rangle$ where R is assumed symmetric, non empty and irriflexive.

2.2.1 Decision-making with Arguments

In this section we simplify part of the content in [10, Chap. 15]. Solving a decision problem amounts to defining a pre-ordering, usually a complete one, on a set $D =$

Fig. 1 An example of AAF

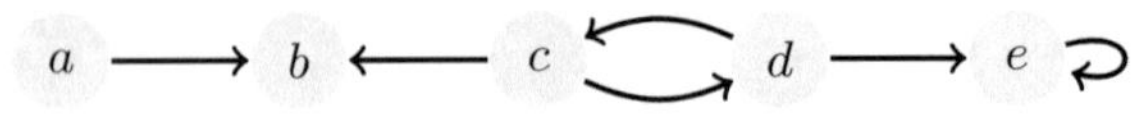

$\{d_1, \ldots, d_n\}$ of n candidate options. Argumentation can be a means for ordering this set D, that is to define a preference relation $\succcurlyeq$ on D. An argumentation-based decision process can be decomposed into the following steps:

1. Constructing arguments in favour/against statements (beliefs or decisions).
2. Evaluating the strength of each argument.
3. Determining the different conflicts among arguments.
4. Evaluating the acceptability of arguments.
5. Comparing decisions on the basis of relevant accepted arguments.

We need to characterise the subsets of practical arguments that are respectively in favour ($\mathcal{F}_f$), or against ($\mathcal{F}_c$) a given option in $d_i \in D$:

- $\mathcal{F}_f : D \to 2^A$ is a function that returns the arguments in favour of a candidate decision. Such arguments are said pros the option.
- $\mathcal{F}_c : D \to 2^A$ is a function that returns the arguments against a candidate decision. Such arguments are said cons the option.

In Definition 5 we present one of the possible ways to prefer ($\succcurlyeq$) one decision instead of another. This *unipolar* principle only refers to either the arguments pros or cons.

Definition 5 *(Counting arguments pros/cons)* Let $DS = (D, F)$ be a decision system, where F is an AAF, and $Acc_{stb}(F)$ collects the sceptically accepted arguments of a framework F under the stable semantics. Let $d_1, d_2 \in D$.

$$d_1 \succcurlyeq d_2 \iff |\mathcal{F}_f(d_1) \cap Acc_{stb}(F)| \geq |\mathcal{F}_f(d_2) \cap Acc_{stb}(F)|$$

The aim of (part of) future work (see also Sect. 5) is to apply similar techniques to derive the best decision about our model, e.g., an evaluation about the software.

3 Proposed Model for MEMCDM Using Argumentation Frameworks

Here, we describe our model: given an MEMCDM problem with n criteria and p experts, how do we "translate"/model it as an AAF? In other words, which arguments and attacks should compose it? Note that, through this section we will use letters S and R to identify "Software", "Ranking" (unlikely to Sect. 2.2, where these letters represent a subset of arguments and the attack relation respectively).

3.1   Arguments

3.1.1   What Does the Data We Use (i.e., Experts' Evaluation of Software in This Case) Tell Us About the Arguments to Add to the Network?

We differentiate arguments that come from the data (i.e., Expert i said that Software j is good) from arguments that are implicit (i.e., Software k is Poor).

1. **Expert i gives Item j a total quality** D_{ij} (which, in the case of Software Quality Assessment – SQA, can be Bad, Poor, Fair, Good, or Excellent):

$$\textbf{Argument}(E_i, S_j, D_{ij})$$

   Let us call such arguments, arguments of type ESD.
2. **Expert i judges that Item j satisfies criterion m up to quality** D_{ijm}

$$\textbf{Argument}(E_i, S_j, c_m, D_{ijm})$$

   Let us call such arguments, arguments of type EScD.

3.1.2   Which Implicit Arguments Should Be Part of the Argumentation Network for This Specific Type of Problem?

For each item, independently from what experts say, there will be a decision made. This decision will be in the form of a final ranking, ranging over all possibly ranking values (in the case of SQA: Bad, Poor, Fair, Good, Excellent). So regardless of ESD arguments, we add to the argumentation network the following arguments:

$$\forall \text{item } S_i, \forall \text{ ranking } D_j : \textbf{Argument } (S_i, D_j)$$

Let us call such arguments, **arguments of type SD**.

3.1.3   Coalitions of Arguments

Here we aim to model the fact the n decisions of any expert on the n criteria of the problem at hand belong together: they together form the "support" for the expert's final decision on the given item. As a result, for any expert E_i and any item S_j, we define a coalition of "supporting" decisions as:

$$\forall E_i, \forall S_j, \textbf{ Coalition: } \{(E_i, S_j, c_k, D_{i,j,k}), \ k \in \{1, \dots, n\}\}$$

Let us call such coalitions of EScDs, extended arguments of type **CoEScD**. The result of modeling such coalitions is that all arguments in the coalition will be forced to be altogether either in or out of extensions. *Per se, we are enforcing an equality constraint on the belonging of these arguments to any extension.*

In the following, in particular in Sects. 3.2.3 and 3.2.4, we will use "*a* supports *b*"as a shortcut to represent a set of attacks from *a* to all the arguments similar, but at teh same time in contrast with *b*.

3.2 Attacks

In this subsection, we answer the following question: What are the **attacks** (*edges of the network*) between these arguments (*nodes*)? *Note:* All attacks we define are reciprocal, hence the edges are always set bidirectionally.

For attacks too, we differentiate between attacks that come from inconsistencies in the decision data (disagreement between experts, inconsistency in decisions of a single expert, lack of fairness, irrationality). An assumption that we make in designing the network model is that experts should be rational: in this, we mean that even if they are not (which we know), they should be and we aim to elicit decisions that are as rational as can be.

3.2.1 Attacks Derived from Lack of Fairness

Here, we assume that if an expert is fair, then s/he should derive the same final ranking from the same criteria rankings. For instance, if there are 3 criteria (c_1, c_2, and c_3) to assess items and an expert E has the following decision history:

$$\begin{cases} E, S_i, c_1, D_1 \\ E, S_i, c_2, D_2 \\ E, S_i, c_3, D_3 \end{cases} \longrightarrow E, S_i, D$$

and: (with $S_i \neq S_j$)

$$\begin{cases} E, S_j, c_1, D_1 \\ E, S_j, c_2, D_2 \\ E, S_j, c_3, D_3 \end{cases} \longrightarrow E, S_j, D'$$

where $D \neq D'$, then we should see arguments (E, S_i, D) and (E, S_j, D') are a lack of fairness in judgment and therefore add the following attack in the argumentation network: $(E, S_i, D) \longleftrightarrow (E, S_j, D')$.

More generally, assuming that the criteria that are considered by the experts are c_k, with $k \in K$, and that the possible rankings are denoted by D_r, with $r \in R$, then we add the following rule to our model:

$$\forall E, S_i, S_j, \quad \text{s.t. } i \neq j \text{ and } \forall k \in K, \ (E, S_i, c_k, D_k) \text{ and } (E, S_j, c_k, D_k):$$

$$\text{if } (E, S_i, D_i) \text{ and } (E, S_j, D_j) \text{ and } D_i \neq D_j$$
$$\text{then} \mathbf{Attack}(E, S_i, D_i) \longleftrightarrow (E, S_j, D_j)$$

3.2.2 Attacks Derived from Lack of Rationality

Let us recall that we assume that the rankings D_r, with $r \in R$, are totally ordered. However, with n criteria, the set of n-tuples of rankings is only partially ordered:

$$(D_1, D_2, \ldots, D_n) \prec (D'_1, D'_2, \ldots, D'_n)$$
$$\text{iff}:$$
$$\forall i \in \{1, \ldots, n\}: \ (D_i \neq D'_i) \longrightarrow D_i < D'_i$$

Now: $\forall E_i$ and $\forall S_j$, we denote by $(D_{1,i,j}, \ldots, D_{n,i,j})$ the set of n decisions made by Expert E_i on each of the criteria $c_1, \ldots, c_n$ for Item S_j, and by $D_{i,j}$ the final decision of Expert E_i on Item S_j.

Being rational for any given expert E_i means that if for Item S_j, s/he ranks criteria lower (w.r.t. above partial order) than s/he ranks the criteria of Item S_k, then his/her final ranking of S_j should not be higher than his/her ranking of S_k. Formally, it is expressed as follows:

$$\forall E_i, \ \forall S_j, \ \forall S_k(j \neq k):$$
$$\text{if: } (D_{1,i,j}, \ldots, D_{n,i,j}) \prec (D_{1,i,k}, \ldots, D_{n,i,k}) \text{ and: } D_{i,j} > D_{i,k}$$
$$\text{then: } \mathbf{Attack} \ (S_j, E_i, D_{i,j}) \longleftrightarrow (S_k, E_i, D_{i,k})$$

3.2.3 Attack Related to Implicit Arguments: SD and cD

In this subsection, we describe the following attacks:

- attacks between implicit arguments SD, and
- attacks across SD and ESD.

1. *Attacks among SDs*: SD Arguments associate an item with a ranking. For each item S_i, there are p SD arguments if there are p possible ranking levels. Each of these p arguments attack each other (they form a complete subgraph). In other words:

$$\forall S_i, \forall r_1, r_2 \in R, \text{ with } r_1 \neq r_2, \ \mathbf{Attack:} \ (S_i, D_{r_1}) \longleftrightarrow (S_i, D_{r_2})$$

2. *Supports between SDs and ESDs*: For any given item S_j, an argument saying that S_i is evaluated D_h is in contradiction (and therefore attacks—and vice-versa) with any argument (E_i, S_j, D_k) as soon as $D_h \neq D_k$. As a result we have:

$$\forall E_i, \ \forall S_j, \ \forall D_k.(E_i, S_j, D_k) \ \textbf{Supports} \ (S_j, D_k)$$

In terms of the attack relation, this support can be rephrased as:

$$\forall E_i, \ \forall S_j, \ (D_h \neq D_k) \ \rightarrow \ \textbf{Attack:} \ (S_j, D_h) \longleftrightarrow (E_i, S_j, D_k)$$

3.2.4 Supports Between Coalitions and ESDs

Here we aim to model the fact that coalitions of decisions on criteria support experts' decisions. In other words:

$$\forall E_i, \forall S_j, \ \{(E_i, S_j, c_k, D_{i,j,k}), \ k \in \{1, \ldots, n\}\} \ \textbf{Supports} \ (E_i, S_j, D_{i,j})$$

In terms of attacks, this is expressed[1] as follows:

$$\forall E_i, E_j \forall S_k : \ D_{i,k} \neq D_{j,k} \ \rightarrow$$
$$\textbf{Attack:} \ \{(E_i, S_k, c_l, D_{i,k,l}), \ k \in \{1, \ldots, n\}\} \longleftrightarrow (E_j, S_k, D_{j,k})$$

4 A Simple Example

Here, let us look at a scenario in which experts independently assess given pieces of software, based on several given evaluation criteria. We describe the resulting argumentation networks (arguments/nodes and attack/edges). Table 1 summarises our example, by reporting all the Poor/Fair/Good quality-evaluation about two different criteria (1 and 2) and the overall quality related to three different software products (S1/S2/S3). Such scores are produced by three different experts (E1/E2/E3). For instance, E1 estimates that the overall quality of S1 is fair, with Criterion 1 evaluated as poor, and Criterion 2 as good.

The graph in Fig. 2 represents the AAF given by following the model proposed in Sect. 3 on the data in Table 1. The yellow nodes represent explicit arguments from the data. The green nodes are the implicit arguments. The blue nodes are the coalitions. The black bold lines represent attacks due to lack of fairness and lack of rationality. The dotted line attacks are those based on implicit arguments. Finally, the grey bold lines are coalition supports of expert decisions.

---

[1]Refer to Sect. 3.1.3 for the definition of support.

Table 1 The explicit arguments that can be collected on our toy-example

Software	Expert	Criterion 1	Criterion 2	Total quality
S1	E1	Poor	Good	Fair
S1	E2	Good	Poor	Fair
S1	E3	Fair	Fair	Fair
S2	E1	Poor	Good	Poor
S2	E2	Poor	Good	Good
S2	E3	Poor	Good	Fair
S3	E1	Good	Good	Good
S3	E2	Good	Good	Fair
S3	E3	Fair	Fair	Fair

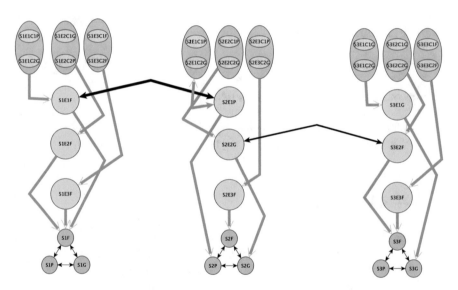

Fig. 2 The AAF given by the model proposed in Sect. 3 on the data in Table 1

4.1 Towards Decision Making

In order to provide a solution to the example in Fig. 2, we adopted a tool developed
by some of the authors of this work, i.e., *ConArg* (*Argumentation with Constraints*).

ConArg[2] [2, 3] is a reasoner based on the *Java Constraint Programming* solver[3]
(JaCoP), a Java library that provides a *Finite Domain Constraint Programming* paradigm [11]. The tool comes with a graphical interface, which visually shows all the
obtained extensions for each problem. ConArg is able to solve also the weighted and

[2]http://www.dmi.unipg.it/conarg/.

[3]http://www.jacop.eu.

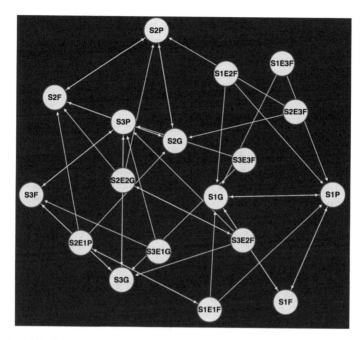

Fig. 3 The AAF of the model in Sect. 3 visualised in the Web interface of ConArg. Arguments containing criteria have been removed from the picture in order simplify it

coalition-based problems presented in [1, 4]. Moreover, it can import/export AAFs with the same text format of ASPARTIX. Recently, we have extended the tool to its second version, i.e., ConArg2 (freely downloadable from the same Web-page of ConArg), in order to improve its performance: we implemented all the models in Gecode,[4] which is an open, free, and efficient C++ environment where to develop constraint-based applications. We have also dropped the graphical interface, having a textual output only. In addition, a Web interface is available at the project homepage, through which it is possible to test the tool without downloading it. So far, ConArg2 finds all conflict-free, admissible, complete, stable, grounded, preferred, semi-stable and ideal extensions (see Definition 3).

In Fig. 3 we visualise the example in Table 1 (and Fig. 2). ConArg returns four stable extensions:

{S1E1F S1E2F S1E3F **S1F** S2E3F **S2F** S3E1G S3E2F S3E3F}
{S1E1F S1E2F S1E3F **S1F** S2E2G S2E3F S3E1G S3E3F}
{S1E2F S1E3F **S1F** S2E1P S2E3F S3E1G S3E2F S3E3F}
{S1E2F S1E3F **S1F** S2E1P S2E2G S2E3F S3E1G S3E3F}

Therefore, we can see that S1F is sceptically accepted (it is present in all the stable extensions), while S2F is credulously accepted (it belongs to the first extension only).

---

[4]http://www.gecode.org.

In addition, no information is obtained for S3, in none of the stable extensions. For this reason, we can safely evaluate S1 as Fair, and, with a lesser degree of certainty, S2 as Fair. With such few experts however, we cannot have a final evaluation for S3.

5   Conclusion and Future Work

In this work, we proposed a model for MEMCDM problems, based on classical AAFs [7], that allows to emulate fairness and rationality. This allows discrimination among input decision data (from experts' prior decisions) between data of value and data that should just not be taken into account. Next steps include operationalising the whole process (from input processing to results filtering) and then adding weights to the attacks to simulate the extent of disagreements and allow lineance towards small errors (e.g., unfairness / irrationality that are really minimal, minor disagreements). Furthermore, we will take inspiration from classical decision-making techniques [10, Ch. 15] with the purpose to rank decisions and decide, for instance, if a software is good or poor. We will even develop new techniques exploring weights on attacks. Also part of future work, we plan to explicitly acknowledge in the AAF that disagreement can be at two different levels: epistemic and pragmatic, and to make use of argumentation frameworks to identify disagreement configurations (epistemic and pragmatic, epistemic only, pragmatic only).

Acknowledgements S. Bistarelli was partially supported by MIUR-PRIN "Metodi logici per il trattamento dell'informazione". M. Ceberio's work was partially supported by the National Science Foundation, NSF CCF grant 0953339 and the American Association for the Advancement of Science, AAAS MIRC (agreement date 112612). F. Santini was partially supported by MIUR PRIN "Security Horizons".

References

1. Bistarelli, S., Santini. F.: A common computational framework for semiring-based argumentation systems. In: ECAI 2010—19th European Conference on Artificial Intelligence, vol. **215**. Frontiers in Artificial Intelligence and Applications, pp. 131–136. IOS Press, (2010)
2. Bistarelli, S., Santini. F.: Conarg: A constraint-based computational framework for argumentation systems. IEEE Computer Society. In: Proceedings of the 2011 IEEE 23rd International Conference on Tools with Artificial Intelligence, ICTAI '11, Washington, DC, USA . pp. 605–612 (2011)
3. Bistarelli, S., Santini. F.: Modeling and solving afs with a constraint-based tool: Conarg. In: Theory and Applications of Formal Argumentation (TAFA). Lecture Notes in Computer Science. vol. **7132**, pp. 99–116. Springer, Heidelberg (2012)
4. Bistarelli, S., Santini. F.: Coalitions of arguments: An approach with constraint programming. Fundam. Inform. **124**(4), 383–401 (2013)
5. Ceberio, M., Modave., F.: An interval-valued, 2-additive Choquet integral for multi-criteria decision making. In: Proceedings of the 10th Conference on Information Processing and Management of Uncertainty in Knowledge-Based Systems (IPMU'04)

6. Coste-Marquis, S., Devred, C., Marquis. P.: Symmetric argumentation frameworks. In: Lluis Godo, (ed.) ECSQARU. Lecture Notes in Computer science, vol. **3571**, pp. 317–328. Springer, (2005)
7. Dung, P.M.: On the acceptability of arguments and its fundamental role in nonmonotonic reasoning, logic programming and n-person games. Artif. Intell. **77**(2), 321–357 (1995)
8. Grabisch, M., Labreuche, C.: A decade of application of the choquet and sugeno integrals in multi-criteria decision aid. 4OR, **6**(1):1–44 (2008)
9. Modave, F., Ceberio, M., Kreinovich, V.: Choquet integrals and OWA criteria as a natural (and optimal) next step after linear aggregation: A new general justification. In: Proceedings of MICAI'2008, pp. 741–753 (2008)
10. Rahwan, I., Simari, G.R.: Argumentation in Artificial Intelligence, 1st edn. Springer Publishing Company, Incorporated, New York (2009)
11. Rossi, F., van Beek, P., Walsh, T.: Handbook of Constraint Programming (Foundations of Artificial Intelligence). Elsevier Science Inc., New York, NY, USA (2006)
12. Wang, X., Ceberio, M., Virani, S., Garcia, A., Cummins, J.: A hybrid algorithm to extract fuzzy measures for software quality assessment. J. Uncertain Syst. **7**(3), 219–237 (2013)
13. Wang, X., Cummins, J., Ceberio, M.: The Bees algorithm to extract fuzzy measures for sample data. In: Proceedings of Annual Conference of North American Fuzzy Information Processing Society (NAFIPS'2011), El Paso, TX, March 2011
14. Wang, X., Garcia Contreras, A.F., Ceberio, M., Del Hoyo, C., Gutierrez, L.C., Virani, S.: Interval-based algorithms to extract fuzzy measures for software quality assessment. In: Proceedings of Annual Conference of North American Fuzzy Information Processing Society (NAFIPS'2012)

Constraint Approach to Multi-objective Optimization

Martine Ceberio, Olga Kosheleva and Vladik Kreinovich

1 Formulation of the Problem

Multi-objective optimization: examples. In many practical situations, we would like to maximize several different criteria.

For example, in meteorology and *environmental research*, it is important to measure fluxes of heat, water, carbon dioxide, methane and other trace gases that are exchanged within the atmospheric boundary layer. To perform these measurements, researchers build up vertical towers equipped with sensors at different heights; these tower are called *Eddy flux* towers. When selecting a location for the Eddy flux tower, we have several criteria to satisfy; see, e.g., [1, 5]: The station should located as far away as possible from roads, so that the gas flux generated by the cars do not influence our measurements of atmospheric fluxes. On the other hand, the station should be located as close to the road as possible, so as to minimize the cost of carrying the heavy parts when building such a station. The inclination at the station location should be small, because otherwise, the flux will be mostly determined by this inclination and will not be reflective of the atmospheric processes, etc.

In *geophysics*, different type of data provide complementary information about the Earth structures. For example, information from the body waves (P-wave receiver functions) mostly covers deep areas, while the information about the Earth surface is mostly contained in surface waves. To get a good understanding of the Earth structure, it is therefore important to take into account data of different types; see, e.g., [3, 9].

If we had only one type of data, then we can use the usual Least Squares approach $f_i(x) \to$ min to find a model that best fits the data. If we knew the relative accuracy

M. Ceberio (✉) · O. Kosheleva · V. Kreinovich
University of Texas at El Paso, El Paso, TX 79968, USA
e-mail: mceberio@utep.edu

O. Kosheleva
e-mail: olgak@utep.edu

V. Kreinovich
e-mail: vladik@utep.edu

© Springer International Publishing AG 2018
M. Ceberio and V. Kreinovich (eds.), *Constraint Programming and Decision Making: Theory and Applications*, Studies in Systems, Decision and Control 100, DOI 10.1007/978-3-319-61753-4_3

of different data types, we could apply the Least Squares approach to all the data. In practice, however, we do not have a good information about the relative accuracy of different data types. In this situation, all we can say that we want to minimize the errors $f_i(x)$ corresponding to all the observations i.

Multi-objective optimization is difficult. The difficulty with this problem is that, in contrast to a simple optimization, the problem of multi-objective optimization is not precisely defined. Indeed, if we want to minimize a single objective $f(x) \to$ min, this has a very precise meaning: we want to find an alternative x_0 for which $f(x_0) \le f(x)$ for all other alternatives x. Similarly, if we want to maximize a single objective $f(x) \to$ max, this has a very precise meaning: we want to find an alternative x_0 for which $f(x_0) \ge f(x)$ for all other alternatives x.

In contrast, for a multi-objective optimization problem

$$f_1(x) \to \text{min}; \quad f_2(x) \to \text{min}; \quad \ldots; \quad f_n(x) \to \text{min} \tag{1}$$

or

$$f_1(x) \to \text{max}; \quad f_2(x) \to \text{max}; \quad \ldots; \quad f_n(x) \to \text{max}, \tag{2}$$

no such precise meaning is known.

Let us illustrate this ambiguity on the above trip example. In many cases, convenient direct flights which save on travel time are more expensive, while a cheaper trip may involve a long stay-over in between flights. So, if we find a trip that minimizes cost, the trip takes longer. Vice versa, if we minimize the travel time, the trip costs more.

It is therefore necessary to come up with a way to find an appropriate compromise between several objectives.

2   Analysis of the Problem and Two Main Ideas

Analysis of the problem. Without losing generality, let us consider a multi-objective maximization problem. In this problem, ideally, we would like to find an alternative x_0 that satisfies the constraints $f_i(x_0) \ge f_i(x)$ for all objectives i and for all alternatives x. In other words, in the ideal case, if we select an alternative x at random, then with probability 1, we satisfy the above constraint.

Main ideas. The problem is that we cannot satisfy all these constraints with probability 1. A natural idea is thus to find x_0 for which the probability of satisfying these constraints is as high as possible. Let us describe two approaches to formulating this idea (i.e., the corresponding probability) is precise terms.

First approach: probability to satisfy all n constraints. The first approach is to look for the probability that for a randomly selected alternative x, we have $f_i(x_0) \ge f_i(x)$ *for all* i.

Second approach: probability to satisfy a randomly selected constraint. An alternative approach is to look for the probability that for a randomly selected alternative x and *for a randomly selected objective* i, we have $f_i(x_0) \geq f_i(x)$.

How to formulate these two ideas in precise terms. To formulate the above two ideas in precise terms, we need to estimate two probabilities:

- the probability $p_I(x_0)$ that for a randomly selected x, we have $f_i(x_0) \geq f_i(x)$ for all i, and
- the probability $p_{II}(x_0)$ that for a randomly selected x and a randomly selected i, we have $f_i(x_0) \geq f_i(x)$.

Let us estimate the first probability. Since we do not have any prior information about the dependence between different objective functions $f_i(x)$ and $f_j(x), i \neq j$, it is reasonable to assume that the events $f_i(x_0) \geq f_i(x)$ and $f_j(x_0) \geq f_j(x)$ are independent for different i and j. Thus, the desired probability $p_I(x_0)$ that all n such inequalities are satisfied can be estimated as the product $p_I(x_0) = \prod_{i=1}^{n} p_i(x_0)$ of n probabilities p_i of satisfying the corresponding inequalities.

So, to estimate p, it is sufficient to estimate, for every i, the probability $p_i(x_0)$ that $f_i(x_0) \geq f_i(x)$ for a randomly selected alternative x.

How can we estimate this probability $p_i(x_0)$? Again, in general, we do not have much prior knowledge of the i-th objective function $f_i(x)$. What do we know? Before starting to solve this problem as a multi-objective optimization problem, we probably tried to simply optimize each of the objective functions—hoping that the corresponding solution would also optimize all other objective functions. Since we are interesting in maximizing, this means that we know the largest possible value M_i of each of the objective functions: $M_i = \max_x f_i(x)$.

In many practical cases, the optimum can be attained by differentiating the objective function and equating all its derivatives to 0. This is, for example, how the Least Squares method works: to optimize the quadratic function that describes how well the model fits the data, we solve the system of linear equations obtained by equating all partial derivatives to 0. It is important to mention that when we consider the points where all the partial derivatives are equal to 0, we find not only maxima but also minima of the objective function. Thus, it is reasonable to assume that in the process of maximizing each objective function $f_i(x)$, in addition to this function's maximum, we also compute its minimum $m_i = \min_x f_i(x)$.

Since we know the smallest possible value m_i of the objective function $f_i(x)$, and we know its largest possible value M_i, we thus know that the value $f_i(x)$ corresponding to a randomly selected alternative x must lie inside the interval $[m_i, M_i]$.

In effect, this is all the information that we have: that the random value $f_i(x)$ is somewhere in the interval $[m_i, M_i]$. Since we do not have any reason to believe that some values from this interval are more probable and some values are less probable, it is reasonable to assume that all the values from this interval are equally probable, i.e., that we have a uniform distribution on the interval $[m_i, M_i]$.

This argument—known as Laplace Indeterminacy Principle—can be formalized as selecting the distribution with the probability density $\rho(x)$ for which the entropy $S = -\int \rho(x) \cdot \ln(\rho(x)) \, dx$ is the largest possible. One can check that for distributions on the given interval, the uniform distribution is the one with the largest entropy [6].

For the uniform distribution on the values $f_i(x) \in [m_i, M_i]$, the probability $p_i(x_0)$ that the random value $f_i(x)$ does not exceed $f_i(x_0)$, i.e., belongs to the subinterval $[m_i, f_i(x_0)]$, is equal to the ratio of the corresponding intervals, i.e., to $p_i(x_0) = \dfrac{f_i(x_0) - m_i}{M_i - m_i}$. Thus, the desired probability $p_I(x_0)$ is equal to the product of such probabilities. So, we arrive at the following precise formulation of the first idea:

Precise formulation of the first idea. To solve a multi-objective optimization problem (2), we find a value x_0 for which the product $p_I(x_0) = \prod\limits_{i=1}^{n} \dfrac{f_i(x_0) - m_i}{M_i - m_i}$ attains the largest possible value, where $m_i \overset{\text{def}}{=} \min\limits_{x} f_i(x)$ and $M_i \overset{\text{def}}{=} \max\limits_{x} f_i(x)$.

Let us estimate the second probability. In the second approach, we select the objective function f_i at random. Since we have no reason to prefer one of the n objective functions, it makes sense to select each of these n functions with equal probability $\dfrac{1}{n}$.

For each selection of the objective function i, we know the probability $p_i(x_0) = \dfrac{f_i(x_0) - m_i}{M_i - m_i}$ that we will have $f_i(x_0) \geq f_i(x)$ for a randomly selected alternative x. The probability of selecting each objective function $f_i(x)$ is equal to $\dfrac{1}{n}$. Thus, we can use the complete probability formula to compute the desired probability $p_{II}(x_0)$:

Precise formulation of the second idea. To solve a multi-objective optimization problem (2), we find a value x_0 for which the expression $p_{II}(x_0) = \sum\limits_{i=1}^{n} \dfrac{1}{n} \cdot \dfrac{f_i(x_0) - m_i}{M_i - m_i}$ attains the largest possible value.

Discussion. Let us show that both ideas lead to known (and widely used) methods for solving multi-objective optimization problems.

The second idea leads to optimizing a linear combination of objective functions. Let us start with analyzing the second idea, since the resulting formula with the sum looks somewhat simpler than the product-based formula corresponding to the first idea.

In the case of the second idea, the optimized value $p_{II}(x_0)$ is a linear combination of n objective functions—to be more precise, it is an arithmetic average of the objective functions normalized in such a way that their values are within the interval $[0, 1]$: $p_{II}(x_0) = \dfrac{1}{n} \cdot \sum\limits_{i=1}^{n} f_i'(x_0)$, where $f_i'(x_0) \overset{\text{def}}{=} \dfrac{f_i(x_0) - m_i}{M_i - m_i}$.

Maximizing a linear combination of the objective functions is indeed the most widely used approach to solving multi-objective optimization problems; see, e.g., [4].

The first idea leads to maximizing a product of (normalized) objective functions. One can easily see that the first idea leads to maximizing a product of normalized objective functions: $p_I(x_0) = \prod_{i=1}^{n} f_i'(x_0)$.

Maximizing such a product is exactly what Bellman-Zadeh fuzzy approach recommends (if we use the product as an "and" operation); see, e.g., [2, 8]. It fits will with our own proposal for such a situation; see, e.g., [5].

This is also exactly what the the Nobelist John Nash recommended for a similar situation of making a group decision when each participant would like to optimize his/her own utility $f_i(x) \to$ max; see, e.g., [7].

Acknowledgements This work was supported in part by the National Science Foundation grants HRD-0734825 and HRD-1242122 (Cyber-ShARE Center of Excellence) and DUE-0926721.

References

1. Aubinet, M., Vesala, T., Papale, D. (eds.): Eddy Covariance - A Practical Guide to Measurement and Data Analysis. Springer, Dordrecht (2012)
2. Bellman, R.E., Zadeh, L.A.: "Decision making in a fuzzy environment". Manag. Sci. **17**(4), B141–B164 (1970)
3. Bodin, T., Sambridge, M., Tkalcic, H., Arroucau, P., Gallagher, K., Rawlinson, N.: "Transdimensional inversion of receiver functions and surface wave dispersion". J. Geophys. Res. Vol. **117**, doi:10.1029/2011JB008560 (2012)
4. Figueira, J., Greco, S., Ehrgott, M. (eds.): Multiple Criteria Decision Analysis: State of the Art Surveys. Kluwer, Dordrecht (2004)
5. Jaimes, A., Tweedie, C., Kreinovich, V., Ceberio, M.: Scale-invariant approach to multi-criterion optimization under uncertainty, with applications to optimal sensor placement, in particular, to sensor placement in environmental research. Int. J. Reliab. Saf. **6**(1–3), 188–203 (2012)
6. Jaynes, E.T., Bretthorst, G.L.: Probability Theory: The Logic of Science. Cambridge University Press, Cambridge, UK (2003)
7. Nash, J.: "Two-Person Cooperative Games". Econometrica **21**, 128–140 (1953)
8. Nguyen, H.T., Walker, E.A.: A First Course in Fuzzy Logic. Chapman and Hall/CRC, Boca Raton, Florida (2006)
9. Thompson, L., Velasco, A., Kreinovich, V.: "Construction of ShearWave models by applying multi-objective optimization to multiple geophysical data sets". In: Tost, G.O., Vasilieva, O. (eds.) Analysis, Modelling, Optimization, and Numerical Techniques. Springer, Berlin (2015, to appear)

From Global to Local Constraints: A Constructive Version of Bloch's Principle

Martine Ceberio, Olga Kosheleva and Vladik Kreinovich

1 Bloch's Principle: Formulation of the Problem

Bloch's Principle: a brief history (see [4] for details). Liouville's Theorem states that every analytical function $f(z)$ which is bounded on a whole complex plane and for which $f(0) = 0$ is equal to 0; see, e.g., [3]. This theorem requires that the constraint $|f(z)| \le C$ be satisfied *globally*, i.e., for all z. What if this constraint is only satisfied *locally*, i.e., for all z from a bounded set? Such a "localization" of Liouville's theorem was indeed proven by H. A. Schwarz: if a function $f(z)$ for which $f(0) = 0$ is analytical for all z from a disk

$$B_R(0) \overset{\text{def}}{=} \{z : |z| < R\}$$

and $|f(z)| \le C$ for all $z \in B_R(0)$, then for all such values z, we get $|f(z)| \le \dfrac{C}{R} \cdot |z|$. When the size R increases, the bound tends to 0; so for $R \to \infty$, we get Liouville's Theorem.

Several similar localizations of global results are known. In 1926, A. Bloch, formulated a general (informal) *Bloch's Principle*: that for every global result, there is a local version from which this global result follows [2]. In complex analysis, this principle was formalized; however, there are many interesting results of the use of Bloch's Principle in other areas of mathematics.

Problem. Can we formalize Bloch's Principle in a context which is more general than complex analysis? If yes, and if the appropriate the localization always exists, can we find it algorithmically?

M. Ceberio (✉) · O. Kosheleva · V. Kreinovich
University of Texas at El Paso, El Paso, TX 79968, USA
e-mail: mceberio@utep.edu

O. Kosheleva
e-mail: olgak@utep.edu

V. Kreinovich
e-mail: vladik@utep.edu

© Springer International Publishing AG 2018
M. Ceberio and V. Kreinovich (eds.), *Constraint Programming and Decision Making: Theory and Applications*, Studies in Systems, Decision and Control 100, DOI 10.1007/978-3-319-61753-4_4

What we do in this paper. In this paper, we provide positive answers to both questions.

Comment. Of course, due to the informal character of Bloch's principle, no answer is final—it is always possible that our result (or a similar result) holds in a more general context.

2   Bloch's Principle: General Formalization

Analysis of the problem. In general terms, the Liouville's theorem has the form

$$\forall f \in \mathcal{F} \, (\forall x \, (f(x) \in A(x)) \Rightarrow \forall x \, (f(x) \in B(x))), \tag{1}$$

where $\mathcal{F}$ is the class of all analytical functions for which $f(0) = 0$, x goes over all complex numbers, $A(x) = \{x : |x| \leq C\}$ is the set of all the valued bounded by the given constant C, and $B(x) = \{0\}$.

The implication (1) says that if the constraint $f(x) \in A(x)$ is *exactly* satisfied for all possible values x, then the conclusion holds. What we want to prove is that when the constraint is "approximately" satisfied—i.e., if it satisfied with some accuracy $\delta > 0$ for all the values x which are at a distance r form 0—then the conclusion is also approximately satisfied, with some accuracy $\varepsilon > 0$ and for all values at a distance R from 0. We also want to make sure that when $\delta \to 0$ and $r \to \infty$, then $\varepsilon \to 0$ and $R \to \infty$. In other words, we want to prove that for every $\varepsilon > 0$ and $R > 0$, there exist $\delta > 0$ and $r > 0$ for which, if the condition is satisfied with accuracy δ for all x which are r-close to 0, then the conclusion is satisfied with accuracy ε for all x which are R-close to 0.

A natural way to describe the fact that $f(x)$ is "approximately" in the set $A(x)$ (or in the set $B(x)$) is to say that $f(x)$ is close to the set $A(x)$ in the sense of the usual distance $d(z, S) \stackrel{\text{def}}{=} \inf\{d(z, s) : s \in S\}$. In the above case, the sets $A(x)$ and $B(x)$ are compact, so $d(z, A(x)) = 0$ if and only if $z \in A(x)$. Thus, the global result (1) can be reformulated in the equivalent form

$$\forall f \, (\forall x \, d(f(x), A(x)) = 0 \Rightarrow \forall x \, d(f(x), B(x)) = 0)). \tag{2}$$

and the desired localized result has the form

$$\forall \varepsilon > 0 \, \forall R > 0 \, \exists \delta > 0 \, \exists r > 0$$

$$\forall f \, ((\forall x \, (d(x, x_0) \leq r \Rightarrow d(f(x), A(x)) \leq \delta)) \Rightarrow$$

$$(\forall x \, (d(x, x_0) \leq R \Rightarrow d(f(x), B(x)) \leq \varepsilon))). \tag{3}$$

It is worth mentioning that in the case of Liouville's Theorem (and in several similar results mentioned in [4]), not only all the sets $A(x)$ and $B(x)$ compact, but they also continuously depend on x – in the sense of the Hausdorff metric $d_H(A, B) \overset{\text{def}}{=}$ $\max\left(\max\limits_{a \in A} d(a, B), \max\limits_{b \in B} d(b, A)\right)$.

The class $\mathcal{F}$ is also compact in some reasonable sense: indeed, for ever bounded set D, the set of all these functions limited to D is compact in the usual metric $d_D(f, g) = \max\limits_{x \in D} d(f(x), g(x))$. Indeed, for an analytical function $f(z)$, its value $f(z)$ can be described by a Cauchy integral over a surrounding curve γ: $f(z) = \int_\gamma \frac{f(t)}{z-t}\, dt$. Differentiation of this formula enables us to get a similar formula for the derivative $f'(z)$. Thus, when the analytical function is bounded, its derivative is also bounded. Due to Ascoli-Arzela theorem, this implies that the corresponding class of functions is compact – when limited to each bounded domain.

It is also important to notice that the notion of an analytical function is *locally defined*, in the sense that if a function $f(x)$ coincides with some analytical function in every neighborhood, then it is analytical itself.

Thus, we arrive at the following natural formalization of Bloch's Principle.

Definition 1 Let us call a metric space *bounded-compact* if every closed bounded set in this space is compact.

Comment. In particular, this implies that every closed ball

$$B_r(x_0) \overset{\text{def}}{=} \{x : d(x, x_0) \leq r\}$$

is compact. Vice versa, if for some point x_0, every closed ball with a center at x_0 is compact then every closed bounded set is compact too: indeed, very bounded set is contained in some ball $B_r(x_0)$, and a closed subset of a compact set is also compact.

Definition 2 Let $\mathcal{F}$ be a class of functions from a bounded-compact metric space X to a bounded-compact metric space Y. We say that the class $\mathcal{F}$ is *bounded-compact* if for every compact set $K \subset X$, this class is compact in the metric $d_K(f, g) \overset{\text{def}}{=} \sup\limits_{x \in K} d(f(x), g(x))$.

Definition 3 Let $\mathcal{F}$ be a class of a functions from X to Y, and let x_0 be a point in x_0. We say that a function $f : X \to Y$ locally belongs to the class $\mathcal{F}$ if for every n, there exists a function $f_n \in \mathcal{F}$ which coincides with f on $B_n(x_0)$.

Comment. This definition uses the point x_0, but one can easily check that the resulting notion does not depend on x_0.

Definition 4 We say that a bounded-compact class of functions $\mathcal{F}$ is *locally defined* if it contains all the functions that locally belong to this class.

Definition 5 Let $\mathcal{F}$ be a bounded-compact locally defined class of functions. By an $\mathcal{F}$-*constraint A*, we mean a (Hausdorff)-continuous function that map each point $x \in X$ into a compact set $A(x) \subseteq Y$.

Definition 6 Let $\mathcal{F}$ be a bounded-compact locally defined class of functions, and let A and B be $\mathcal{F}$-constraints.

- We say that the constraint A *globally implies* the constraint B if for every function $f \in \mathcal{F}$, the condition $\forall x \, (f(x) \in A(x))$ implies $\forall x \, (f(x) \in B(x))$.
- We say that the constraint A *locally implies* the constraint B for ε, R, δ, and r if for every function $f(x)$ for which $d(f(x), A(x)) \leq \delta$ for all x with $d(x, x_0) \leq r$, we have $d(f(x), B(x)) \leq \varepsilon$ for all x with $d(x, x_0) \leq R$.
- We say that the constraint A *locally implies* the constraint B if for every $\varepsilon > 0$ and for every $R > 0$, there exist real numbers $\delta > 0$ and $r > 0$ such that A locally implies B for ε, R, δ, and r.

Comment. The definition of local implication uses a point x_0, but one can easily see that the corresponding property does not change if we replace this point with any other point from the metric spaces X.

Proposition 1 *Let $\mathcal{F}$ be a bounded-compact locally defined class of functions, and let A and B are $\mathcal{F}$-constraints. Then, if A globally implies B, then A locally implies B.*

Proof We will prove the result by contradiction. Let us assume that A does not locally imply B. This means that there exist $\varepsilon > 0$ and $R > 0$ such that for every n, there is a function $f_n \in \mathcal{F}$ for which $\max_{x \in B_n(x_0)} d(f_n(x), A(x)) \leq 1/n$ but $d(f_n(x_n), B(x_n)) > \varepsilon$ for some $x_n \in B_{x_0}(R)$. Since the sequence x_n is contained in a compact set $B_R(x_0)$, it has a subsequence which converges to some limit ℓ. Without losing generality, we can assume that $x_n \to \ell$.

Since $\mathcal{F}$ is compact relative to each metric $d_{B_k(x_0)}$, from the sequence f_n, we can extract a subsequence $n(1, i)$ convergent for $k = 1$; from this subsequence, we can extract a subsequence $n(2, i)$ which is convergent for $k = 2$, etc. The diagonal subsequence $f_{n(i,i)}$ then converges for all k. This convergence is for all x, no matter how far from x_0 we are, so we can defining a point-wise limit function $f(x)$. On each ball $B_k(x_0)$, this limit coincides with the corresponding limit from $\mathcal{F}$ limited to this ball. Thus, the limit function $f(x)$ locally belongs to $\mathcal{F}$; since the class $\mathcal{F}$ is locally defined, this means that $f \in \mathcal{F}$.

For the limit function f, for every x, the condition $d(f_n(x), A(x)) \leq 1/n$ in the limit tends to $d(f(x), A(x)) = 0$. Since A globally implies B, we conclude that we have $d(f(x), B(x)) = 0$ for all x, in particular, that we have $d(f(\ell), B(\ell)) = 0$. However, from $d(f_n(x_n), B(x_n)) > \varepsilon$, in the limit $x_n \to \ell$, we get $d(f(\ell), B(\ell)) \geq \varepsilon > 0$. This contradictions shows that our assumption is wrong, and A does locally imply B. The proposition is proven.

3 Bloch's Principle: A Constructive Version

Towards an algorithmic version. In this paper, we will use the usual definitions of computable numbers, functions, compact spaces, etc.; see, e.g., [1, 5].

Proposition 2 *If spaces X and Y are computable and computably bounded-compact, and if A and B are computable functions for which A globally implies B, then there exists an algorithm that, given rational numbers $\varepsilon > 0$ and $R > 0$, produces computable numbers $\delta > 0$ and $r > 0$ for which A locally implies B for ε, R, δ, and r.*

Proof From the proof of Proposition 1 can conclude that that for $\varepsilon_0 = \varepsilon/3$ and for $R_0 = R + 1$, there exists an integer $n = n_0$ for which $r = n$ and $\delta = 1/n$ satisfy the desired property. Let us show how to algorithmically find this n. For that, we will repeat the below computations for $n = 1, 2, \ldots$ until we find the value n for which the desired condition is satisfied.

In these computations, we will use the fact that there are algorithms for computing the maximum and minimum of a computable function over a computable compact. We will also use the fact that for a computable function $F(x)$ on a computable compact set K, for every two computable numbers $z^- < z^+$ within the range of $F(x)$ on K, we can compute an intermediate value $z \in (z^-, z^+)$ for which the set $\{x : F(x) \leq z\}$ is a computable compact.

Before we go through $n = 1, 2, \ldots$, we use the intermediate-value algorithm to compute a value $R' \in (R, R + 1)$ for which the ball $B_{R'}(x_0)$ is computably compact.

Then, for each n, we compute a value $r_n \in (n - 1, n)$ for which the closed ball $B_{r_n}(x_0)$ is a computable compact. Since this ball is a computable compact, the value $v(f) \overset{\text{def}}{=} \max_{x \in B_{r_n}(x_0)} d(f(x), A(x))$ is also computable – and is, therefore, a computable function of $f \in \mathcal{F}' \overset{\text{def}}{=} \mathcal{F}_{|B_{x_0}(R)}$.

The restriction $\mathcal{F}'$ is a computable compact. Thus, by the same intermediate-value result, we can compute a value $\delta_n \in (1/n, 1/(n-1))$ for which the set $S \overset{\text{def}}{=} \{f : v(f) \leq \delta_n\}$ is a computable compact. We can therefore compute the maximum M of a computable function $d(f(x), B(x))$ over all $x \in B_{R'}(x_0)$ and all $f \in S$ with any given accuracy. Let us compute it with accuracy $\varepsilon/3$. If the resulting estimate $\widetilde{M}$ is $\leq (2/3) \cdot \varepsilon$, we stop.

Let us show that if we stop, then we get the desired n. Indeed, in this case, if for some f, we have $d(f(x), A(x)) \leq 1/n < \delta_n$ for all $x \in B_n(x_0)$, then (since $r_n < n$) this inequality is also true for all $x \in B_{r_n}(x_0)$, hence $v(f) < \delta_n$. Every $x \in B_R(x_0)$ belongs to $B_{R'}(x_0)$ and thus, for this x, we have $d(f(x), B(x)) \leq M$. Since $M \leq \widetilde{M} + \varepsilon/3$ and $\widetilde{M} \leq (2/3) \cdot \varepsilon$, we conclude that $d(f(x), B(x)) \leq \varepsilon$.

Let us now show that the above algorithm will stop for $n = n_0 + 1$. By definition of n_0, if $x \in B_{n_0}(x_0)$ and $d(f(x), A(x)) \leq 1/n_0$, then $d(f(x), B(x)) \leq \varepsilon/3$ for all $x \in B_{R_0}(x_0)$. Here, $R' < R + 1 = R_0$, so $x \in B_{R'}(x_0)$ implies that $x \in B_{R_0}(x_0)$. Similarly, since $r_n > n - 1 = n_0$, we conclude that $\max_{x \in B_{n_0}(x_0)} d(f(x), A(x)) \leq v(f) =$

$\max_{x \in B_{r_n}(x_0)} d(f(x), A(x))$ and thus, $v(f) \leq \delta_n < 1/n_0$ implies that $\max_{x \in B_{n_0}(x_0)} d(f(x),$

$A(x)) < \dfrac{1}{n_0}$. Thus, indeed, for all such x and f, we have $d(f(x), B(x)) \leq \varepsilon/3$;

hence, the largest value M is $\leq \varepsilon/3$, so $\widetilde{M} \leq (2/3) \cdot \varepsilon$, and the algorithm will stop. The proposition is proven.

Acknowledgements This work was supported in part by the National Science Foundation grants HRD-0734825, HRD-124212, and DUE-0926721.

References

1. Bishop, E., Bridges, D.S.: Constructive Analysis. Springer, New York (1985)
2. Bloch, A.: "La conception actuelle de la theorie de fonctions entieres et meromorphes". Enseignement mathematique **25**, 83–103 (1926)
3. Lang, S.: Complex Analysis. Springer, New York (2003)
4. Osserman, R.: From Schwarz to Pick to Ahlfors and beyond. Not. Am. Math. Soc. **46**(8), 868–873 (1999)
5. Weihrauch, K.: Computable Analysis. Springer, Berlin (2000)

Optimizing pred(25) Is NP-Hard

Martine Ceberio, Olga Kosheleva and Vladik Kreinovich

1 Formulation of the Problem

Need to estimate parameters of models. In many practical situations, we know that a quantity y depends on the quantities $x_1, \ldots, x_n$, and we know the general type of this dependence. In precise terms, this means that we know a family of functions $f(c_1, \ldots, c_p, x_1, \ldots, x_n)$ characterized by parameters c_i, and we know that the actual dependence corresponds to one of these functions.

For example, we may know that the dependence is linear; in this cases, the corresponding family takes the form

$$f(c_1, \ldots, c_n, c_{n+1}, x_1, \ldots, x_n) = c_{n+1} + \sum_{i=1}^{n} c_i \cdot x_i.$$

In general, we know the type of the dependence, but we do not know the actual values of the parameters. These values can only be determined from the measurements and observations, when we observe the values x_j and the corresponding value y. Measurement and observations are always approximate, so we end up with tuples $(x_{1k}, \ldots, x_{nk}, y_k)$, $1 \le k \le K$, for which $y_k \approx f(c_1, \ldots, c_p, x_{1k}, \ldots, x_{nk})$ for all k. We need to estimate the parameters $c_1, \ldots, c_p$ based on these measurement results.

Least Squares: traditional way of estimating parameters of models. In most practical situations, the Least Squares method is used to estimate the desired parameters. In this method, we select the values c_i for which the sum of the squares of the approximation errors is the smallest possible:

M. Ceberio (✉) · O. Kosheleva · V. Kreinovich
University of Texas at El Paso, El Paso, TX 79968, USA
e-mail: mceberio@utep.edu

O. Kosheleva
e-mail: olgak@utep.edu

V. Kreinovich
e-mail: vladik@utep.edu

© Springer International Publishing AG 2018
M. Ceberio and V. Kreinovich (eds.), *Constraint Programming and Decision Making: Theory and Applications*, Studies in Systems, Decision and Control 100, DOI 10.1007/978-3-319-61753-4_5

$$\sum_k (y_k - f(c_1, \ldots, c_p, x_{1k}, \ldots, x_{nk}))^2 \to \min_{c_1, \ldots, c_p} .$$

One of advantages of this approach is that, when the model $f(c_1, \ldots, c_p, x_1, \ldots, x_n)$ linearly depends on the parameters c_i, the sum of squares is a quadratic function of c_i. Thus, when we apply the usual criterion for the minimum—differentiate the sum with respect to each variable x_i and equate all the resulting partial derivatives to 0—we get a system of linear equations, from which we can easily find all the unknown $c_1, \ldots, c_p$.

Least Squares is not always the optimal way of estimating the parameters. The Least Squares approach known to be optimal for the case when all the approximation errors $y_k - f(c_1, \ldots, c_p, x_{1k}, \ldots, x_{nk})$ are independent and all distributed according to the same normal distribution. In practice, however, we often have outliers—e.g., values corresponding to the malfunction of a measuring instrument—and in the presence of even a single outlier, the Least Squares method can give very wrong results.

Let us illustrate this on the simplified example, when y does not depend on any variables x_i at all, i.e., when $y = c$ for some unknown constant c. In this case, we need to estimate the value c based on the observations $y_1, \ldots, y_K$. For this problem, the Least Squares method takes the form $\sum_{k=1}^{K} (y_k - c)^2 \to \min$. Differentiating the sum with respect to the unknown c and equating the derivative to 0, we conclude that $c = \dfrac{y_1 + \ldots + y_K}{K}$.

This formula works well if all the values y_i are approximately equal to c. For example, if the actual value of c is 0, and $|y_i| \leq 0.1$, we get an estimate $|c| \leq 0.1$. However, if out of 100 measurements y_i, one of an outlier equal to 1000, the estimate becomes close to 10—and thus, far away from the actual value 0.

To take care of such situations, we need estimates which do not change as much in the presence of possible outliers. Such methods are called *robust* [2].

pred(25) as an example of a robust estimate. One of the possible robust estimates consists of selecting a percentage α and selecting the values of the parameters for which the number of observations for which the prediction is within $\alpha\%$ from the observed value is the largest possible. In other words, each prediction is formulated as a constraint, and we look for parameters that maximize the number of satisfied constraint. This technique is known as pred(α).

This method is especially widely used in software engineering, e.g., for estimating how well different models can predict the overall software effort and/or the number of bugs. In software engineering, this method is most frequently applied as pred(25), for $\alpha = 25$; see, e.g., [1, 3].

Problem. In contrast to the Least Squares approach, for which the usual calculus ideas lead to an efficient optimization algorithm, no such easy solution is known for pred(25) estimates; all known algorithms for this estimation are rather time-consuming. A natural question arises: is this because we have not yet found a feasible algorithm for computing these estimates, or is this estimation problem really hard?

What we prove in this paper. In this paper, we prove that even for a linear model with no free term c_{n+1}, pred(25) estimation—as well as pred(α) estimation for any $\alpha > 0$—is an NP-hard problem. In plain terms, this means that this problem is indeed inherently hard.

2 Main Result and Its Proof

Definition 1 Let $\alpha \in (0, 1)$ be a rational number. By a *linear pred(α)-estimation problem*, we means the following problem:

- *Given:* an integer n, K rational-valued tuples $(x_{1k}, \ldots, x_{nk}, y_k)$, $1 \le k \le K$, and an integer $M < K$;
- *Check:* whether there exist parameters $c_1, \ldots, c_n$ for which in at least M cases k, we have

$$\left| y_k - \sum_{i=1}^{n} c_i \cdot x_{ik} \right| \le \alpha \cdot \left| \sum_{i=1}^{n} c_i \cdot x_{ik} \right|.$$

Proposition 1 *For every α, the linear pred(α)-estimation problem is NP-hard.*

Proof To prove this result, we will reduce, to this problem, a known NP-hard problem of checking whether a set of integer weights $s_1, \ldots, s_m$ can be divided into two parts of equal overall weight, i.e., whether there exist integers $y_j \in \{-1, 1\}$ for which $\sum_{j=1}^{m} y_j \cdot s_j = 0$; see, e.g., [4].

In the reduced problem, we will have $n = m + 1$, with $n = m + 1$ unknown coefficients $c_1, \ldots, c_m, c_{m+1}$. The parameters c_i will correspond to the values y_i, and c_{m+1} is equal to 1. We will build tuples corresponding to equations $y_i = 1$ and $y_i = -1$ for $i \le m$, to $c_{m+1} = 1$, and to the equation $c_{m+1} + \sum_{i=1}^{m} y_i \cdot s_i = 1$.

To each equation of the type $y_i = 1$ or $c_{m+1} = 1$, we put into correspondence the following two tuples:

- In the first tuple, $x_{ik} = 1 + \varepsilon$, $x_{jk} = 0$ for all $j \ne i$, and $y_k = 1$. The resulting linear term has the form $c_i \cdot (1 + \varepsilon)$ and thus, the corresponding inequality takes the form $1 - \varepsilon \le (1 + \varepsilon) \cdot c_i \le 1 + \varepsilon$, i.e., equivalently, the form $\dfrac{1 - \varepsilon}{1 + \varepsilon} \le c_i \le 1$.
- In the second tuple, $x_{ik} = 1 - \varepsilon$, $x_{jk} = 0$ for all $j \ne i$, and $y_k = 1$. The resulting linear term has the form $c_i \cdot (1 - \varepsilon)$ and thus, the corresponding inequality takes the form $1 - \varepsilon \le (1 - \varepsilon) \cdot c_i \le 1 + \varepsilon$, i.e., equivalently, the form $1 \le c_i \le \dfrac{1 + \varepsilon}{1 - \varepsilon}$.

It should be mentioned that the only value c_i that satisfies both inequalities is the value $c_i = 1$.

Similarly, to each equation of the type $y_i = -1$, we put into correspondence following two tuples.

- In the first tuple, $x_{ik} = 1 + \varepsilon$, $x_{jk} = 0$ for all $j \neq i$, and $y_k = -1$. The resulting linear term has the form $c_i \cdot (1 + \varepsilon)$ and thus, the corresponding inequality takes the form $-1 - \varepsilon \leq (1 + \varepsilon) \cdot c_i \leq -1 - \varepsilon$, i.e., equivalently, the form $-1 \leq c_i \leq -\dfrac{1 - \varepsilon}{1 + \varepsilon}$.

- In the second tuple, $x_{ik} = 1 - \varepsilon$, $x_{jk} = 0$ for all $j \neq i$, and $y_k = -1$. The resulting linear term has the form $c_i \cdot (1 - \varepsilon)$ and thus, the corresponding inequality takes the form $-1 - \varepsilon \leq (1 - \varepsilon) \cdot c_i \leq -1 + \varepsilon$, i.e., equivalently, the form $-\dfrac{1 + \varepsilon}{1 - \varepsilon} \leq c_i \leq -1$.

Here also, the only value c_i that satisfies both inequalities is the value $c_i = -1$.

Finally, to the equation $c_{m+1} + \sum\limits_{j=1}^{m} y_j \cdot s_j = 1$, we put into correspondence the following two tuples. In both tuples, $y_k = 1$.

- In the first tuple, $x_{ik} = (1 + \varepsilon) \cdot s_i$, and $x_{m+1,k} = 1 + \varepsilon$. The corresponding linear term has the form $(1 + \varepsilon) \cdot \left(\sum\limits_{i=1}^{m} c_i \cdot s_i + c_{m+1} \right)$, and thus, the corresponding inequality takes the form

$$1 - \varepsilon \leq (1 + \varepsilon) \cdot \left(\sum\limits_{i=1}^{m} c_i \cdot s_i + c_{m+1} \right) \leq 1 + \varepsilon,$$

i.e., equivalently,

$$\frac{1 - \varepsilon}{1 + \varepsilon} \leq \sum\limits_{i=1}^{m} c_i \cdot s_i + c_{m+1} \leq 1.$$

- In the second tuple, $x_{ik} = (1 - \varepsilon) \cdot s_i$, and $x_{m+1,k} = 1 - \varepsilon$. The corresponding linear term has the form $(1 - \varepsilon) \cdot \left(\sum\limits_{i=1}^{m} c_i \cdot s_i + c_{m+1} \right)$, and thus, the corresponding inequality takes the form

$$1 - \varepsilon \leq (1 - \varepsilon) \cdot \left(\sum\limits_{i=1}^{m} c_i \cdot s_i + c_{m+1} \right) \leq 1 + \varepsilon,$$

i.e., equivalently,

$$1 \leq \sum\limits_{i=1}^{m} c_i \cdot s_i + c_{m+1} \leq \frac{1 + \varepsilon}{1 - \varepsilon}.$$

Here, both inequalities are satisfied if and only if $\sum\limits_{i=1}^{m} c_i \cdot s_i + c_{m+1} = 1$.

Overall, we have $2m + 2$ pairs, i.e., $4m + 4$ tuples. If for the given values $s_1, \ldots, s_m$, the original NP-hard problem has a solution y_i, then we can take $c_i = y_i$,

$c_{m+1} = 1$, and thus satisfy $M \overset{\text{def}}{=} 2m + 4$ inequalities. Let us show that, vice versa, if at least $2m + 4$ inequalities are satisfied, this means that the original problem has a solution.

Indeed, for every i, each of the two inequalities corresponding to $y_i = 1$ implies that $c_i > 0$ while each of the two inequalities corresponding to $y_i = -1$ implies that $c_i < 0$. Thus, these inequalities incompatible, which means that for every i, at most two inequalities can be satisfied. If for some i, fewer than two inequalities are satisfied, then even when for every $j \neq i$, we have two, and all four remaining inequalities are satisfied, we will still have fewer than $2m + 4$ satisfied inequalities. This means that if $2m + 4$ inequalities are satisfied, then for every i, two inequalities are satisfied—and thus, either $c_i = 1$ or $c_i = -1$. Now, the four additional inequalities also have to be satisfied, so we have $c_{m+1} = 1$, and $\sum_{i=1}^{m} c_i \cdot s_i + c_{m+1} = 1$, hence $\sum_{i=1}^{m} c_i \cdot s_i = 0$. The reduction is proven, and thus our problem is indeed NP-hard.

Comment. In this proof, we consider situations in which about half of the inequalities are satisfied. We may want to restrict ourselves to situations in which a certain proportion of inequality should be satisfied—e.g., 90% or 99%. With such a restriction, the problem remains NP-hard.

To prove this, it is sufficient to consider a similar reduction, in which:

- instead of single pair of tuples corresponding to $c_{m+1} = 1$ we have N identical pairs (for a sufficiently large N), and similarly,
- instead of a single pair corresponding to the equation $\sum_{j=1}^{m} y_j \cdot s_j = 0$, we have N such identical pairs.

Acknowledgements This work was supported in part by the National Science Foundation grants HRD-0734825 and HRD-1242122 (Cyber-ShARE Center of Excellence) and DUE-0926721.

References

1. Conte, S.D., Dunsmore, H.E., Shen, V.Y.: Software Engineering Metrics and Models. Benjamin/Cummings, Menlo Park, California (1986)
2. Huber, P.J.: Robust Statistics. Wiley, Hoboken, New Jersey (2004)
3. Mendes, E.: Cost Estimation Techniques for Web Projects. IGI Publisher, Hershey, Pennsylvania (2007)
4. Papadimitriou, C.H.: Computational Complexity. Addison Wesley, San Diego (1994)

Range Estimation Under Constraints Is Computable Unless There Is a Discontinuity

Martine Ceberio, Olga Kosheleva and Vladik Kreinovich

1 Need for Range Estimation Under Constraints

Need for data processing. To make a decision, in particular, to select an engineering design and/or control strategy, we need to know the effects of selecting different alternatives. In most engineering problems, we know how different quantities depend on each other and how they change with time. In particular, we usually know how the quantity y describing the effect depends on the values of the quantities $x_1, \ldots, x_n$ describing the decision and the surrounding environment: $y = f(x_1, \ldots, x_n)$. The resulting computations are known as *data processing*.

    Need to take uncertainty into account. In the ideal situation, when we know the exact values $x_1, \ldots, x_n$ of the corresponding parameters, we can simply substitute these values into a known function f, and get the desired value y. In practice, the values $x_1, \ldots, x_n$ come from measurements, and measurements are never absolutely accurate. As a result, the measurement results $\widetilde{x}_1, \ldots, \widetilde{x}_n$ are, in general, somewhat different from the actual (unknown) values $x_1, \ldots, x_n$ of the corresponding quantity. Thus, the estimate $\widetilde{y} = f(\widetilde{x}_1, \ldots, \widetilde{x}_n)$ is, in general, different from the desired value $y = f(x_1, \ldots, x_n)$. To make an appropriate decision, it is important to know how big can be the difference $\widetilde{y} - y$.

    Need for range estimation. In many practical situations, the only information that we have about the measurement error $\widetilde{x}_i - x_i$ of each corresponding measurements is the upper bound Δ_i provided by the manufacturer. In this case, based on the measurement result $\widetilde{x}_i$, the only information that we can conclude about x_i is that x_i belongs to the *interval* $[\underline{x}_i, \overline{x}_i] \stackrel{\text{def}}{=} [\widetilde{x}_i - \Delta_i, \widetilde{x}_i + \Delta_i]$.

M. Ceberio (✉) · O. Kosheleva · V. Kreinovich
University of Texas at El Paso, El Paso, TX 79968, USA
e-mail: mceberio@utep.edu

O. Kosheleva
e-mail: olgak@utep.edu

V. Kreinovich
e-mail: vladik@utep.edu

© Springer International Publishing AG 2018
M. Ceberio and V. Kreinovich (eds.), *Constraint Programming and Decision Making: Theory and Applications*, Studies in Systems, Decision and Control 100, DOI 10.1007/978-3-319-61753-4_6

Another case of such an interval uncertainty is when the parameter x_i characterizes a manufactured part; in this case, we know that the corresponding value must lie within the tolerance interval—the interval $[\underline{x}_i, \overline{x}_i]$ within which the manufacturer of this part was required to keep this value.

Different values x_i from the corresponding intervals $[\widetilde{x}_i - \Delta_i, \widetilde{x}_i + \Delta_i]$ lead, in general, to different values of $y = f(x_1, \ldots, x_n)$. It is therefore important to estimate the range of all such values, i.e., the set

$$\{f(x_1, \ldots, x_n) : x_i \in [\underline{x}_i, \overline{x}_i] \text{ for all } i\}.$$

In the usual case of continuous functions f, this range is an interval; we will denote this interval by $[\underline{y}, \overline{y}]$. Estimation of this range interval is known as *interval computations*; see, e.g., [4].

Range estimation problems are, in general, computable. It is known that for computable functions f on computable intervals $[\underline{x}_i, \overline{x}_i]$, there is an algorithm which computes the range of the given function on given intervals; see, e.g., [3].

In general, the corresponding computational problem is NP-hard (meaning that these computations may take a very long time), but there are many situations where feasible algorithms are possible for exact computations—and there are also many feasible algorithms for providing enclosures for the desired ranges; see, e.g., [3].

Need to take constraints into account. The above formulation of range estimation problem assumes that the quantities $x_1, \ldots, x_n$ are independent—in the sense that the set of possible values of, e.g., x_1, does not depend on the actual values of all other quantities. In practice, we often have additional *constraints* which limit possible combinations of values $(x_1, \ldots, x_n)$.

For example, if x_1 and x_2 represent the control values are two consequent moments of time, then usually, due to hardware limitations, these values cannot differ much, we should have a constraint $|x_1 - x_2| < \delta$ for some small value $\delta > 0$. In this case, instead of the range of all possible values of $f(x_1, \ldots, x_n)$ when each x_i is in the corresponding interval, we are only interested in the range of the values corresponding to the tuples $(x_1, \ldots, x_n)$ that satisfy all the known constraints.

Constraints make the problem of range estimation more complex. Adding constraints immediately makes the problem much more complex; see, e.g., [1].

What we do in this paper. In this paper, we explain that the main reason why range estimation under constraints is not always computable is that constraints may introduce discontinuity—and all computable functions are continuous. Specifically, we show that if we restrict ourselves to computable continuous constraints, the problem of range estimation under constraints remains computable.

2 Known Results: Brief Reminder

Definition 1

- A real number x is called *computable* if there exists an algorithm that, given a natural number k, returns a rational number r_k for which $|r_k - x| \le 2^{-k}$.
- An interval $[\underline{x}, \overline{x}]$ is called *computable* if both its endpoints are computable.
- A function $f(x_1, \ldots, x_n)$ from real numbers to real numbers is called *computable* if there exist two algorithms:

  - an algorithm that, given rational numbers $r_1, \ldots, r_n$, and an integer k, returns a rational number r for which $|r - f(r_1, \ldots, r_n)| \le 2^{-k}$; and
  - an algorithm that, given a rational number $\varepsilon > 0$, returns a rational number $\delta > 0$ such that if $|x_i - x_i'| \le \delta$ for all i, then

$$|f(x_1, \ldots, x_n) - f(x_1', \ldots, x_n')| \le \varepsilon.$$

Proposition 1 *[3, 5] There exists an algorithm that, given a computable function $f(x_1, \ldots, x_n)$ and computable intervals $[\underline{x}_i, \overline{x}_i]$ $(1 \le i \le n)$, returns the range $[\underline{y}, \overline{y}]$ of this function on the given intervals.*

Proof To compute $\overline{y}$ with a given accuracy $\varepsilon > 0$, we first use the second algorithm from the definition of a computable function to find $\delta > 0$ for which $|x_i - x_i'| \le \delta$ implies that the values of f are $(\varepsilon/2)$-close to each other. On each interval $[\underline{x}_i, \overline{x}_i]$, we then select finitely many points $\underline{x}_i, \underline{x}_i + \delta, \underline{x}_i + 2\delta, \ldots$ After that, for each combination $(s_1, \ldots, s_n)$ of the selected points, we use the first algorithm to produce a rational number r which is $(\varepsilon/2)$-close to the corresponding value $f(s_1, \ldots, s_n)$. Our claim is that the largest $\overline{r}$ of these rational numbers is the desired ε-approximation to $\overline{y}$.

Indeed, on the one hand, each rational value r is bounded by $f(s_1, \ldots, s_n) + \dfrac{\varepsilon}{2}$. Thus, from $f(s_1, \ldots, s_n) \le \overline{y}$, we conclude that $r \le \overline{y} + \dfrac{\varepsilon}{2}$. In particular, this is true for the largest of these numbers, hence $\overline{r} \le \overline{y} + \dfrac{\varepsilon}{2}$.

On the other hand, let us consider the values x_i at which the function f attains its largest possible value $\overline{y}$: $f(x_1, \ldots, x_n) = \overline{y}$. Each value x_i from the corresponding interval is δ-close to one of the selected points s_i. Thus, each combination $(x_1, \ldots, x_n)$ is δ-close to the corresponding combination $(s_1, \ldots, s_n)$ of selected points—which, due to the choice of δ, implies that

$$|f(s_1, \ldots, s_n) - f(x_1, \ldots, x_n)| \le \frac{\varepsilon}{2}.$$

So, $f(s_1, \ldots, s_n) \ge \overline{y} - \dfrac{\varepsilon}{2}$. For the corresponding number r, we have $r \ge f(s_1, \ldots, s_n) - \dfrac{\varepsilon}{2}$ and hence, $r \ge \overline{y} - \varepsilon$. Since $\overline{r}$ is the largest of these rational numbers, we get $\overline{r} \ge r$ and therefore, $\overline{r} \ge \overline{y} - \varepsilon$.

A similar proof shows that the smallest $\underline{r}$ of the corresponding rational numbers r is an ε-approximation to $\underline{y}$. The proposition is proven.

Definition 2

- By a *computable constraint*, we mean a constraint of one of the following types: $g_j(x_1, \ldots, x_n) = c_j$, $g_j(x_1, \ldots, x_n) \leq c_j$, $c_j \leq g_j(x_1, \ldots, x_n)$, or $\underline{c}_j \leq g_j(x_1, \ldots, x_n) \leq \overline{c}_j$, where $g_j(x_1, \ldots, x_n)$ is a computable function and c_j, $\underline{c}_j$, and $\overline{c}_j$ are computable numbers.
- By a problem of *range estimation under constraints*, we mean the following problem:

  - given a computable function $f(x_1, \ldots, x_n)$, n computable intervals $[\underline{x}_i, \overline{x}_i]$, and a finite list of computable constraints,
  - compute the largest $\overline{y}$ and the smallest y values of $f(x_1, \ldots, x_n)$ for all the tuples $(x_1, \ldots, x_n)$ of values $x_i \in [\underline{x}_i, \overline{x}_i]$ which satisfy all the given constraints.

Proposition 2 *No algorithm is possible which solves all the problems of range estimation under constraints.*

Comment In other words, it is not possible to have an algorithm that, given the function f, the intervals $[\underline{x}_i, \overline{x}_i]$, and the constraints, would always compute the values $\underline{y}$ and $\overline{y}$.

Proof Let us take $n = 1$, $f(x_1) = x_1$, and a constraint $g(x_1) = c_1$, where $g(x_1) = \min(x_1, \max(0, x_1 - 1))$. One can check that for $x_1 \leq 0$, we get $g(x_1) = x_1$; for $0 \leq x_1 \leq 1$, we get $g(x_1) = 0$, and for $x_1 \geq 1$, we get $g(x_1) = x_1 - 1$. So, for $c_1 < 0$, the constraint is only satisfied for the value $x_1 = c_1$, so we get $\overline{y} = c_1$; on the other hand, for $c_1 = 0$, the constraint $g(x_1) = c_1 = 0$ is satisfied for all $x_1 \in [0, 1]$, so we get $\overline{y} = 1$. When $c_1 \to 0$, the dependence of $\overline{y}$ on c_1 is discontinuous, and all computable functions are continuous; see, e.g., [5]. The proposition is proven.

3 New Result: Discontinuity Is the only Obstacle to Computing $\underline{Y}$ and $\overline{Y}$

Definition 3

- Let the computable intervals $[\underline{x}_i, \overline{x}_i]$ be given, and let the computable functions $g_1(x_1, \ldots, x_n)$, …be given, and for each of these functions, let a type of the corresponding constraint be given (i.e., $= c_j$, $\leq c_j$, $\geq c_j$, or $\underline{c}_j \leq \cdot \leq \overline{c}_j$).
- For each combination c of the threshold values c_j, $\underline{c}_j$, and/or $\overline{c}_j$, by $S(c)$, we denote the set of all the tuples $x_i \in [\underline{x}_i, \overline{x}_i]$ which satisfy all the corresponding constraints.

- For each $\delta > 0$, we say that the combinations c and c' are δ-close if the corresponding threshold are δ-close (e.g., $|c_j - c'_j| \le \delta$).
- We say that the set of constraints is *computably continuous* if there exists an algorithm that, given a rational number $\varepsilon > 0$, returns a rational number $\delta > 0$ such that when c and c' are δ-close, then $d_H(S(c), S(c')) \le \varepsilon$, where $d_H(A, B)$ is the Hausdorff distance $d_H(A, B) \overset{\text{def}}{=} \max \left(\sup_{a \in A} d(a, B), \sup_{b \in B} d(b, A) \right)$ and $d(a, B) \overset{\text{def}}{=} \inf_{b \in B} d(a, b)$.

Proposition 3 *There exists an algorithm which solves the problem of range estimation under constraints for all computably continuous constraints.*

Comment In other words, this algorithm, given the function $f(x_1, \ldots, x_n)$, the intervals $[\underline{x}_i, \overline{x}_i]$, and the constraints, returns the corresponding values $\underline{y}$ and $\overline{y}$.

Proof To estimate $\underline{y}$ and $\overline{y}$ with accuracy ε, let us find $\delta > 0$ for which $|x_i - x'_i| \le \delta$ implies that the f-values are ε-close. One can then show that if $d_H(S, S') \le \delta$, then $\max_{x \in S} f(x)$ and $\max_{x \in S'} f(x)$ are ε-close [2].

For this $\delta > 0$, we can find $\beta > 0$ for which if c and c' are β-close, then $d_H(S(c), S(c')) \le \delta$. We can now replace each equality $g_j = c_j$ with inequalities $\underline{c}_j \le g_j \le \overline{c}_j$ and, as long as $|\underline{c}_j - c_j| \le \beta$ and $|\overline{c}_j - c_j| \le \beta$, we still have a δ-close set $S(c)$. The box $[\underline{x}_1, \overline{x}_1] \times \ldots$ is a computable compact set (see [1, 3, 5]), so due to the known properties of such sets, there exists β-close values c' for which the set $S(c')$ is a computable compact—and for which, therefore, the maximum $\overline{y}'$ and the minimum $\underline{y}'$ of the computable function $f(x)$ over $S(c')$ are computable. Since $S(c')$ is δ-close to $S(c)$, we have $|\overline{y}' - \overline{y}| \le \varepsilon$ and $|\underline{y}' - \underline{y}| \le \varepsilon$. The proposition is proven.

Proposition 4 *When all constraints are inequalities, with $\underline{c}_j < \overline{c}_j$, then we can solve all problems of range estimation for which the dependence $S(c)$ is continuous (not necessarily computably continuous).*

Proof For $\beta = 2^{-k}$, $k = 0, 1, \ldots$, we estimate the ranges $[\underline{y}'_j, \overline{y}'_j]$ and $[\underline{y}''_j, \overline{y}''_j]$ of f over an inner β-approximation $S(c')$ and the outer β-approximations $S(c'')$. Then $\underline{y}'' \le \underline{y} \le \underline{y}'$ (and $\overline{y}' \le \overline{y} \le \overline{y}''$). Due to continuity, the sets $S(c')$ and $S(c'')$ will eventually become δ-close and thus, the estimates $\underline{y}'$ and $\underline{y}''$ become ε-close; when this happens, we return $\underline{y}'$ and $\overline{y}'$ as the desired ε-approximations to $\underline{y}$ and $\overline{y}$.

Acknowledgements This work was supported in part by the National Science Foundation grants HRD-0734825, HRD-124212, and DUE-0926721.

References

1. Ceberio, M., Kreinovich, V. (eds.): Constraint Programming and Decision Making. Springer, Heidelberg (2014)
2. Kreinovich, V., Kubica, B.: From computing sets of optima, Pareto sets, and sets of Nash equilibria to general decision-related set computations. J. Univers. Comput. Sci. **16**(18), 2657–2685 (2010)
3. Kreinovich, V., Lakeyev, A., Rohn, J., Kahl, P.: Computational Complexity And Feasibility Of Data Processing And Interval Computations. Kluwer, Dordrecht (1997)
4. Moore, R. E., Kearfott, R. B., Cloud, M. J.: Introduction to Interval Analysis, SIAM Press (2009)
5. Weihrauch, K.: Computable Analysis. Springer, Berlin (2000)

Towards a Physically Meaningful Definition of Computable Discontinuous and Multi-valued Functions (Constraints)

Martine Ceberio, Olga Kosheleva and Vladik Kreinovich

1 Formulation of the Problem

Need to define computable discontinuous functions. One of the main objectives of physics it to predict physical phenomena, i.e., use the observations to compute the predicted values of the corresponding physical quantities. Many physical phenomena such as phase transitions and quantum transitions include discontinuous dependencies $y = f(x)$ ("jumps"); see, e.g., [2].

In other physical situations, for some values x, we may have several possible values y. From the purely mathematical viewpoint, this means that the relation between x and y is no longer a function, it is a *relation* of a *constraint* $R \subseteq X \times Y$; following the terminology widely used in applications, we will also call them *multi-valued functions*.

To analyze which models of discontinuous or multi-valued behavior are computable and which are not, we need to have a precise definition of what is means for a discontinuous and/or multi-valued function to be computable. Alas, the current definitions of computable functions are mostly limited to continuous case.

What we plan to do. Our main goal is to define what it means for a discontinuous and/or multi-valued function to be computable.

For that purpose, we first explain the current definitions of computable numbers, objects, and functions. Then, we use physical motivations to come up with a new definition of computable discontinuous and multi-valued functions. Finally, we provide a few preliminary results about the new definition.

Computable numbers: reminder. Intuitively, a real number is *computable* if we can compute it with any desired accuracy. In more precise terms, a real number x is

M. Ceberio (✉) · O. Kosheleva · V. Kreinovich
University of Texas at El Paso, El Paso, TX 79968, USA
e-mail: mceberio@utep.edu

O. Kosheleva
e-mail: olgak@utep.edu

V. Kreinovich
e-mail: vladik@utep.edu

© Springer International Publishing AG 2018
M. Ceberio and V. Kreinovich (eds.), *Constraint Programming and Decision Making: Theory and Applications*, Studies in Systems, Decision and Control 100, DOI 10.1007/978-3-319-61753-4_7

called *computable* if there exists an algorithm that, given a natural number n, returns a rational number r_n which is 2^{-n}-close to x: $|x - r_n| \leq 2^{-n}$; [1, 3].

Computable metric spaces: motivation. A similar notion of computable elements can be defined for general metric spaces. In general, a element x is computable if there is an algorithm which generates better and better approximation to x. At each moment of time, we only have a finite amount of information about x; based on this information, we produce an approximation corresponding to this information. Any information can be represented, in the computer, as a sequence of 0s and 1s; any such sequence can be, in turn, interpreted as a binary integer n. Let $\tilde{x}_n$ denote an approximation corresponding to an integer n. Then, it makes sense to require that in a computable metric space, there is a sequence of such approximating elements $\{\tilde{x}_n\}$.

Computable means, in particular, that the distance $d_X(\tilde{x}_n, \tilde{x}_m)$ between such elements should be computable. Thus, we arrive at the following definition.

Computable metric spaces: definition. By a computable metric space, we mean a metric space X with a sequence $\{\tilde{x}_n\}$ of elements such that there is an algorithm that, given two natural numbers m and n, returns the distance $d_X(\tilde{x}_m, \tilde{x}_n)$ (i.e., for every natural number k, returns a rational number r_k which is 2^{-k}-close to $d_X(\tilde{x}_m, \tilde{x}_n)$)).

We say that an element x of a computable metric space X is *computable* if there exists an algorithm that, given a natural number n, returns an integer k_n for which $\tilde{x}_{k_n}$ is 2^{-n}-close to x: $d_X(\tilde{x}_{k_n}, x) \leq 2^{-n}$.

Computable functions: definition. A function $f : X \to Y$ from a computable metric space X to a computable metric space Y is called *computable* if there exists an algorithm which uses x as an input and computes, for each integer n, a 2^{-n}-approximation y_k to $f(x)$. By "uses x as an input", we mean that to compute y_k, this algorithm can request, for each integer m, a 2^{-m}-approximation x_ℓ to x (and to use the index ℓ of this 2^{-m}-approximation in computing y_k).

Computable functions are continuous. The problem with the above definition is that all the functions computable according to this definition are continuous; see, e.g., [1, 3]. Thus, we cannot use this definition to check how well we can compute a discontinuous function.

This continuity is easy to understand. For example, if we have a function $f(x)$ form real numbers to real numbers which is equal to 0 for $x \leq 0$ an to 1 for $x > 0$, then, if we could compute $f(x)$ for a given x with accuracy 2^{-2}, then we would be able, given a computable real number x, to tell whether this number is positive or not, and this is known to be algorithmically impossible.

Computable compact set. In analyzing computability, it is often useful to start with *pre-compact* metric spaces, i.e., metric spaces X for which, for every positive real number $\varepsilon > 0$, there exists a finite ε-net, i.e., a finite list of elements L such that every element $x \in X$ is ε-close to one of the elements from this list. In a Euclidean space, every bounded set is compact. A pre-compact set is *compact* if every converging sequence has a limit.

A natural idea is to call a compact metric space X *computable compact* if X is a computable metric space and there is an (additional) algorithm that, given an integer n, returns a finite list L_n of elements of X which is a 2^{-n}-net for X.

2 Towards a New Definition of Computable Discontinuous and Multi-valued Functions

Simplifying comment. Before we start analyzing the problem, let us make one important comment. Functions can not only be discontinuous or multi-valued, they can also be undefined for some inputs x. However, in contrast to discontinuity and multiplicity of values, this is not a serious problem: if a relation is not everywhere defined, we can make it everywhere defined if we consider, instead of the original set X, a projection of R on this set. For example, a function $\sqrt{x}$ is not everywhere defined on the real line, but it is everywhere defined on the set of all non-negative real numbers. Thus, without losing generality, we can assume that our relation is everywhere defined.

Definition 1 A relation $R \subseteq X \times Y$ is called *everywhere defined* if for every $x \in X$, there exists a $y \in Y$ for which $(x, y) \in R$.

Analysis of the problem. From the physical viewpoint, what does it mean that the dependence between x and y—as described by a given discontinuous and/or multi-valued function—is computable?

In the ideal case, when we have a continuous single-valued dependence, the value x uniquely determines the value $y = f(x)$. In this case, once we know x, we want to compute $f(x)$ with a given accuracy. This is exactly the idea behind the usual definition of a computable function.

For a multi-valued function, for the same input x, we may get several different values y. In this case, it is desirable to compute the *set* of all possible value y corresponding to a given x. When we limit ourselves to multi-valued mappings from a compact set X to a compact set Y, the set of x-possible values of y is pre-compact, and thus, with any given accuracy, can be described by a finite list L of possible values. In other words:

- first, the list L should represent *all* possible values, i.e., if y is a possible value of $f(x)$ for a given x, then y should be close to one of the values from the finite list L;
- second, all the values from the list L must be possible values; in other words, for every value from the list, there must exist a close possible value of $f(x)$.

Discontinuity provides an additional complexity which can be illustrated on the example of the above discontinuous function $f(x) = 0$ for $x \leq 0$ and $f(x) = 1$ for $x > 0$. In particular, for $x = 0$, we get $f(x) = f(0) = 0$. However, at each stage of the computation, we only know an approximate value of x. So, when the actual value of the input is $x = 0$, we will never find out whether x is non-positive (in which case $f(x) = 0$) or positive (in which case $f(x) = 1$). Thus, no matter how accurately we measure x, the only information about y that we can conclude is y is either equal to 0 or equal to 1. In general, we need to take into account not only the values $f(x)$ for a given x, but also the values $f(x')$ corresponding to values x' which are close to x. In view of this, the above properties of the list L must be appropriately modified:

- first, the list L should represent *all* possible values, i.e., if y is a possible value of $f(x')$ for some x' which is close to the given x, then y should be close to one of the values from the finite list L;
- second, all the values from the list L must be possible values; in other words, for every value from the list, there must exist a close value y which is a possible value of $f(x')$ for some x' which is close to x.

In general, the closeness does does not have to be the same in both cases. Thus, we arrive at the following definition.

Definition 2 Let X and Y be computable compact sets with metrics d_X and d_Y. An everywhere defined relation $R \subseteq X \times Y$ is called *computable* if there exists an algorithm that, given a computable element $x \in X$ and computable positive numbers $0 < \varepsilon < \varepsilon'$ and $0 < \delta$, produces a finite list $\{y_1, \ldots, y_m\} \subseteq Y$ that satisfies the following two properties:

(1) if $(x', y) \in R$ for some x' for which $d_X(x', x) \leq \varepsilon$, then there exists an i for which $d_Y(y, y_i) \leq \delta$;
(2) for each element y_i from this list, there exist values x' and y for which $d_X(x, x') \leq \varepsilon'$, $d_Y(y_i, y) \leq \delta$, and $(x', y) \in R$.

3 Properties of the New Definition

Main result. If X and Y are metric spaces with metrics d_X and d_Y, then on their Cartesian product $X \times Y$ (i.e., the set of all pairs (x, y), $x \in X$ and $y \in Y$) we can define a metric $d_{X \times Y}((x, y), (x', y')) \stackrel{\text{def}}{=} \max(d_X(x, x'), d_Y(y, y'))$. One can check that if X and Y are both compact sets, then the product $X \times Y$ is also a compact set: to get an ε-net for $X \times Y$, it is sufficient to take ε-nets L_X for X and L_Y for Y; one can then easily check that the set $L_X \times L_y$ of all possible pairs is an ε-net for the Cartesian product $X \times Y$. This construction is computable, so we conclude that the Cartesian product of computable compact sets is also a computable compact set.

Our first—somewhat surprising—result is that this new definition is equivalent to simply requiring that the set R (describing the graph of the relation) is a computable compact set:

Proposition 1 *Let X and Y be computable compact sets. A relation $R \subseteq X \times Y$ is computable if and only if the set R is a computable compact set.*

Proof $\Leftarrow$ Let us first prove that if R is a computable compact set, then the relation R is computable in the sense of Definition 2. Indeed, let x be a computable element of X, and let the computable positive values $\varepsilon < \varepsilon'$ be given. Then, according to a known result from [1], we can find a computable value $\varepsilon_0 \in (\varepsilon, \varepsilon')$ for which the set $S \stackrel{\text{def}}{=} \{(x', y) \in R : d_X(x, x') \leq \varepsilon_0\}$ is also a computable compact set. Thus, for a given computable number $\delta > 0$, there exists a finite δ-net for this set S. Let us denote the elements of this δ-net L by $(x_1, y_1), \ldots, (x_m, y_m)$. Let us show that, as the desired finite list, we can now take the list $\{y_1, \ldots, y_m\}$. Let us prove that this list satisfies both desired properties.

(1) If $(x', y) \in R$ for some x' for which $d_X(x, x') \leq \varepsilon$, then, due to $\varepsilon < \varepsilon_0$, we have $d_X(x, x') < \varepsilon_0$. Thus, $(x', y) \in S$. Since $L = \{(x_1, y_1), \ldots, (x_m, y_m)\}$ is a δ-net for the set S, we conclude that there exists an index i for which $d_{X \times Y}((x', y), (x_i, y_i)) \leq \delta$. By definition of $d_{X \times Y}$, this means that $\max(d_X(x', x_i), d_Y(y, y_i)) \leq \delta$ and therefore, $d_Y(y, y_i) \leq \delta$. The first property from Definition 1 is proven.

(2) Let us now prove the second property. Let y_i be one of the selected elements. By our construction, the corresponding pair (x_i, y_i) belongs to δ-net for the set S. In particular, this means that $(x_i, y_i) \in S$. This means that $(x_i, y_i) \in R$ and that $d_X(x, x_i) \leq \varepsilon_0$. Since $\varepsilon_0 < \varepsilon'$, we conclude that $d_X(x, x_i) \leq \varepsilon'$. Thus, for each i, there exists $x' = x_i$ and $y = y_i$ for which $d_X(x, x') \leq \varepsilon'$, $d_Y(y_i, y) = 0 \leq \delta$, and $(x', y) \in R$. The second property is proven as well.

$\Rightarrow$ Let us now prove that if R is a computable relation in the sense of Definition 2, then R is computable compact set. For that, we must show how, given a computable positive real number $\alpha > 0$, we can generate an α-net for this set R. First, we use that fact that X is a computable compact, and generate an $(\alpha/2)$-net $\{x_1, \ldots, x_k\}$. For each point x_i, we then apply Definition 2 for $\delta = \varepsilon = \alpha/2$ and $\varepsilon' = \alpha$ and generate the corresponding list $\{y_{i1}, \ldots, y_{im_i}\}$. Let us show that the pairs (x_i, y_{ij}) form an α-net for the set R.

Indeed, by Definition 2, for each i and j, there exist values x' and y for which $d_X(x_i, x') \leq \varepsilon' = \alpha$, $d_Y(y_{ij}, y) \leq \delta = \alpha/2$, and $(x', y) \in R$. Thus, the pair (x_i, y_{ij}) is α-close to an element $(x', y) \in R$.

Vice versa, let $(x, y) \in R$. Since x_i form an $(\alpha/2)$-net, there exists an i for which $d(x, x_i) \leq \alpha/2 = \varepsilon$. From Property (1) of Definition 2, we can now conclude that there exists a j for which $d_Y(y, y_{ij}) \leq \delta = \alpha$. Thus, $d_{X \times Y}((x, y), (x_i, y_{ij})) = \max(d_X(x, x_i), d_Y(y, y_{ij})) \leq \max(\alpha/2, \alpha) = \alpha$. The proposition is proven.

Inverse functions: a corollary. If the range of R is the whole set Y, then, from Proposition 1, it follows that a multi-valued function (relation) R is computable if and only if its inverse $R^{-1} = \{(x, y) : (y, x) \in R\}$ is computable.

Acknowledgements This work was supported in part by NSF grants HRD-1242122 and DUE-0926721, by NIH Grants 1 T36 GM078000-01 and 1R43TR000173-01, and by an ONR grant N62909-12-1-7039.

References

1. Bishop, E., Bridges, D.S.: Constructive Analysis. Springer, New York (1985)
2. Feynman, R., Leighton, R., Sands, M.: The Feynman Lectures on Physics. Addison Wesley, Boston, Massachusetts (2005)
3. Weihrauch, K.: Computable Analysis. Springer, Berlin (2000)

Algebraic Product is the only T-norm for Which Optimization Under Fuzzy Constraints is Scale-Invariant

Juan Carlos Figueroa-García, Martine Ceberio and Vladik Kreinovich

1 Formulation of the Problem

Need for optimization under fuzzy constraints. In decision making, we would like to find the best solution x among all possible solutions.

For example, if we need to build a chemical plant for producing chemicals needed for space exploration and for sophisticated electronics, then we need to select a design which is the most profitable among all the designs whose possible negative effect on the environment is small. In this example, the objective function is the overall profit.

In this example (and in many similar examples) the objective functions is well defined in the sense that for each alternative x, we can compute the exact value $f(x)$ of the objective function for this particular design. In contrast, the constraints are *not* well-defined, they are formulated by using words from a natural language (like "small"), words which are nor precise.

A reasonable way to describe the meaning of such imprecise ("fuzzy") constraints is to use techniques of fuzzy logic (see, e.g., [4, 6, 8]), where to each possible alternative x, we assign a number $\mu_c(x)$ describing to what extent this design satisfies the corresponding constraint. To find this value $\mu_c(x)$, we can, e.g., ask the user to mark this extent on a scale from 0 to 10, and if the user marks 7, take $\mu_c(x) = 7/10$.

This way, the original problem becomes a problem of optimization under fuzzy constraint: find x for which $f(x)$ is the largest possible among all x which satisfy the constraint described by a function $\mu_c(x)$.

Bellman-Zadeh approach to optimization under fuzzy constraints. To solve such problems, R. Bellman (a known specialist in optimization) and L. Zadeh (the founder of the fuzzy logic approach) came back with the following idea; see, e.g., [1, 4].

J.C. Figueroa-García (✉)
Universidad Distrital Francisco José de Caldas, Bogotá, Colombia
e-mail: filthed@gmail.com

M. Ceberio · V. Kreinovich
Department of Computer Science, University of Texas at El Paso, El Paso, TX 79968, USA
e-mail: mceberio@utep.edu

V. Kreinovich
e-mail: vladik@utep.edu

© Springer International Publishing AG 2018
M. Ceberio and V. Kreinovich (eds.), *Constraint Programming and Decision Making: Theory and Applications*, Studies in Systems, Decision and Control 100,
DOI 10.1007/978-3-319-61753-4_8

51

First, we (somehow) find the smallest value m of the objective function $f(x)$ among all possible solutions x, and we also find the largest possible value M of the objective function over all possible constraints. based on the values m and M, we can form, for each alternative x, the degree $\mu_m(x)$ to which x is maximal, as $\mu_m(x) \stackrel{\text{def}}{=} \dfrac{f(x) - m}{M - m}$. The larger $f(x)$, the larger this degree, and it reaches the value 1 if $f(x)$ attains the largest possible value M.

We want to find an alternative which satisfies the constraints *and* optimizes the objective function. In fuzzy techniques, the degree of truth in "and"-statement is approximately described by applying an appropriate *t-norm* $f(a, b)$ to the degrees to which both statements are true; see, e.g., [4, 6]. A t-norm must satisfy several natural properties: e.g., the fact that $A \,\&\, B$ means the same as $B \,\&\, A$ leads to the commutativity $f_\&(a, b) = f_\&(b, a)$, and the fact that "true" $\&\, A$ is equivalent to just A leads to the property $f_\&(1, a) = a$.

- By applying the t-norm $f_\&(a, b)$ to the degrees $\mu_c(x)$ and $\mu_m(x)$, we find the degrees $\mu_s(x) = f_\&(\mu_c(x), \mu_m(x))$ to which each alternative x is a solution.
- We then select the alternative which is the best fit, i.e., for which the degree $\mu_s(x)$ is the largest.

Problem: the value M is not well defined. Usually, we have some prior experience with similar problems, so we know some alternative(s) x which were previously selected. The value $f(x)$ for such "status quo" alternatives can be used as the desired minimum m.

Finding M is much more complicated, we do not know which alternatives to include and which not to include. If we replace the original value M with a new value $M' > M$, then the maximizing degree changes, from $\mu_m(x) = \dfrac{f(x) - m}{M - m}$ to $\mu'_m(x) = \dfrac{f(x) - m}{M' - m}$. One can easily see that $\mu'_m(x) = \lambda \cdot \mu_m(x)$ for $\lambda \stackrel{\text{def}}{=} \dfrac{M - m}{M' - m} < 1$.

The problem is that in general, the alternatives for which the functions $\mu_s(x) = f_\&(\mu_c(x), \mu_m(x))$ and $\mu'_s(x) = f_\&(\mu_c(x), \mu'_m(x)) = f_\&(\mu_c(x), \lambda \cdot \mu_m(x))$ may be different.

It is therefore desirable to come up with a scheme in which the solution would not change if we simply re-scale $\mu_m(x)$ by modifying the not well-defined quantity M.

What we do in this paper. In this paper, we show that the dependence on M disappears if we use algebraic product t-norm $f_\&(a.b) = a \cdot b$. We also show that this is the only t-norm for which decisions do not depend on M.

2   Main Results

Definition 1 *By a t-norm, we mean a function* $f_\&ast : [0, 1] \times [0, 1] \to [0, 1]$ *for which* $f_\&ast(a, b) = f_\&ast(b, a)$ *and* $f_\&ast(1, a) = a$ *for all a and b.*

Comment. Usually, it is also required that the t-norm is associative. However, our results do not need associativity, so they are valid for non-associative and-operations as well; such non-associative operations are sometimes used to more adequately describe human reasoning; see, e.g., [2, 3, 5, 7, 9].

Definition 2 *Let* $f_\&ast(a, b)$ *be a t-norm. We say that optimization under fuzzy constraints is scale-invariant for this t-norm if for every set X, for every two functions* $\mu_c : X \to [0, 1]$ *and* $\mu_m : X \to [0, 1]$, *and for every real number* $\lambda \in (0, 1)$, *we have* $S = S'$, *where:*

- S is the set of all $x \in X$ for which the function $\mu_s(x) = f_\&ast(\mu_c(x), \mu_m(x))$ attains its maximum, i.e., for which $\mu_s(x) = \max_{y \in X} \mu_s(y)$;
- S' is the set of all $x \in X$ for which the function $\mu'_s(x) = f_\&ast(\mu_c(x), \lambda \cdot \mu_m(x))$ attains its maximum, i.e., for which $\mu'_s(x) = \max_{y \in X} \mu'_s(y)$.

Proposition 1 *For the algebraic product t-norm* $f_\&ast(a, b) = a \cdot b$, *optimization under fuzzy constraints is scale-invariant.*

Proposition 2 *The algebraic product t-norm* $f_\&ast(a, b) = a \cdot b$ *is the only t-norm for which optimization under fuzzy constraints is scale-invariant.*

Proof of Proposition 1. For the algebraic product t-norm:

- S is the set of all $x \in X$ for which the function $\mu_s(x) = \mu_c(x) \cdot \mu_m(x)$ attains its maximum, and
- S' is the set of all $x \in X$ for which the function $\mu'_s(x) = \mu_c(x) \cdot \lambda \cdot \mu_m(x)$ attains its maximum.

Here, $\mu'_s(x) = \lambda \cdot \mu_s(x)$ for a positive number λ. Clearly, $\mu_s(x) \geq \mu_s(y)$ if and only if $\lambda \cdot \mu_s(x) \geq \lambda \cdot \mu_s(y)$, so the optimizing sets S and S' indeed coincide.

Proof of Proposition 2. Let $f_\&ast(a, b)$ be a t-norm for which optimization under fuzzy constraints is scale-invariant, and let a and b be two number from the interval $[0, 1]$. Let us prove that $f_\&ast(a, b) = a \cdot b$.

Let us consider $X = \{x_1, x_2\}$ with $\mu_c(x_1) = \mu_m(x_2) = a$ and $\mu_c(x_2) = \mu_m(x_1) = 1$. In this case, $\mu_s(x_1) = f_\&ast(\mu_c(x_1), \mu_m(x_1)) = f_\&ast(a, 1)$. Due to commutativity, we get $\mu_s(x_1) = f_\&ast(1, a)$ and due to the second property of the t-norm, we get $\mu_s(x_1) = a$.

Similarly, we have $\mu_s(x_2) = f_\&ast(\mu_c(x_2), \mu_m(x_2)) = f_\&ast(1, a)$. Due to the second property of the t-norm, we also get $\mu_s(x_2) = a$.

Since $\mu_s(x_1) = \mu_s(x_2)$, the optimizing set S consists of both elements x_1 and x_2.

Due to scale-invariance, for $\lambda = b$, the same set $S' = S = \{x_1, x_2\}$ must be the optimizing set for the function $\mu'_s(x) = f_\&(\mu_c(x), \lambda \cdot \mu_m(x))$. Thus, we must have $\mu'_s(x_1) = \mu'_s(x_2)$, i.e., $f_\&(a, b \cdot 1) = f_\&(1, b \cdot a)$. So, $f_\&(a, b) = f_\&(1, a \cdot b)$. Due to the second property of the t-norm, we conclude that $f_\&(a, b) = a \cdot b$.

The proposition is proven.

Acknowledgements This work was supported in part by the National Science Foundation grants 0953339, HRD-0734825 and HRD-1242122 (Cyber-ShARE Center of Excellence) and DUE-0926721, by Grants 1 T36 GM078000-01 and 1R43TR000173-01 from the National Institutes of Health, and by a grant N62909-12-1-7039 from the Office of Naval Research.

This work was performed when Juan Carlos Figueroa-García was a visiting researcher at the University of Texas at El Paso.

References

1. Bellman, R.E., Zadeh, L.A.: Decision making in a fuzzy environment. Manag. Sci. **47**(4), B141–B145 (1970)
2. Bouchon-Meunier, B., Kreinovich, V., Nguyen, H.T.: "Non-Associative Operations". In: Proceedings of the Second International Conference on Intelligent Technologies InTech'2001, Bangkok, Thailand, November 27–29, pp. 39–46 (2001)
3. Goodman, I.R., Trejo, R.A., Kreinovich, V., Martinez, J., Gonzalez, R.: "An even more realistic (non-associative) interval logic and its relation to psychology of human reasoning", In: Proceedings of the Joint 9th World Congress of the International Fuzzy Systems Association and 20th International Conference of the North American Fuzzy Information Processing Society IFSA/NAFIPS'2001, Vancouver, Canada, July 25–28, pp. 1586–1591 (2001)
4. Klir, G.J., Yuan, B.: Fuzzy Sets and Fuzzy Logic. Prentice Hall, Upper Saddle River, New Jersey (1995)
5. Goodman, I.R., Trejo, R.A., Kreinovich, V., Martinez, J., Gonzalez, R.: "An even more realistic (non-associative) interval logic and its relation to psychology of human reasoning", In: Proceedings of the Joint 9th World Congress of the International Fuzzy Systems Association and 20th International Conference of the North American Fuzzy Information Processing Society IFSA/NAFIPS'2001, Vancouver, Canada, July 25–28, pp. 1586–1591 (2001)
6. Nguyen, H.T., Walker, E.A.: First Course In Fuzzy Logic. CRC Press, Boca Raton, Florida (2006)
7. Trejo, R., Kreinovich, V., Goodman, I.R., Martinez, J., Gonzalez, R.: "A realistic (Non-associative) logic and a possible explanations of 7±2 Law". Int. J. Approx. Reason. **29**, 235–266 (2002)
8. Zadeh, L.A.: Fuzzy sets. Inf. Control. **8**, 338–353 (1965)
9. Zimmerman, H.H., Zysno, P.: Latent connectives in human decision making. Fuzzy Sets Syst. **4**, 37–51 (1980)

Comparing Operation Points in Linear Programming with Fuzzy Constraints

Juan Carlos Figueroa-García, Germán Hernández-Pérez and Dusko Kalenatic

1 Introduction

Optimization over uncertain environments is a challenge for many decision makers who need solving different problems. Some of those problems require the use of special algorithms, so it is an interesting field to be covered. In some cases, there is no any certainty to have stable conditions in the system, so we need to keep in mind that such uncertainty affects decision making and the way to solve the problem.

One of the most popular problems in decision making is Linear Programming (LP) due to its applicability and efficiency. Similarly, Fuzzy Linear Programming (FLP) gained popularity as a powerful technique to handle non probabilistic uncertainty, adding flexibility to classical LP problems. The first fuzzy constrained LP model has been proposed by Zimmermann [15] which is equivalent to the work of Delgado, Verdegay and Vila [5]. Zimmermann and Fullér [16] defined some decision making principle over fuzzy environments. Interval valued optimization has been extensively treated by Hladík [8], Černý and Hladík [4], Fiedler et al. [6], whose results are intimately related to fuzzy optimization.

Other interesting works on FLP has been wrote by Zimmermann and Fullér [16], Mahdavi-Amiri and Nasseri [11], Friedman et al. [7], Ramík [13, 14], Inuiguchi and Ramík [9], Ramík and Řimánek [12], Campos [2], Campos and Verdegay [3], and Rommelfanger [14] who analyzed fuzzy optimization and FLPs from different points of view, and all their contributions has been taken into account in this work.

As fuzzy techniques became popular in 80's and 90's, some important issues came from its implementation. An important practical issue regards to compare optimal

J.C. Figueroa-García (✉)
Universidad Distrital Francisco José de Caldas, Bogotá, Colombia
e-mail: jcfigueroag@udistrital.edu.co

G. Hernández-Pérez
Universidad Nacional de Colombia, Bogotá Campus, Bogotá, Colombia
e-mail: gjhernandezp@gmail.com

D. Kalenatic
Universidad de La Sabana, Chia, Colombia
e-mail: duskokalenatic@yahoo.com

© Springer International Publishing AG 2018
M. Ceberio and V. Kreinovich (eds.), *Constraint Programming and Decision Making: Theory and Applications*, Studies in Systems, Decision and Control 100,
DOI 10.1007/978-3-319-61753-4_9

solutions vs. achieved solutions, which means that even when an optimal solution of a problem is available, what in reality is implemented could be different, so the analyst needs to compare the optimal vs. an implemented solution, and how to move closer to its optimal solution.

The soft constraints method proposed by Zimmermann [15] is based on a symmetrical handling of the linear fuzzy constraints of an LP model, which means that if the goal increases, its possibility of occurrence decreases since more resources are required to increase the goal. This way, we propose a way to quantify the difference between the optimal solution provided by the Zimmermann's method and solutions achieved by different strategies applied in practice.

The chapter is intended to analyze the relationship between theoretical optimal solutions in FLPs (Zimmermann's soft constraints method) and what is obtained in practical applications. This relationship is important in the sense that optimal solutions are not always reachable in practice, so what the analyst can do is compare the achieved results to the theoretical optimal solution in order to see how far/close is it. To do so, we propose a ranking index that compares an obtained solution of an FLP problem to its optimal solution, and a way to compute its membership degree through the Bellman-Zadeh decision making and the Zadeh extension principle.

The chapter is divided into five sections. Section 1 introduces the topic and present some relevant bibliography. Section 2 presents the Zimmermann's soft constraints method. Section 3 shows some concepts about optimality in FLP problems. Section 4 presents the proposed ranking method and an application example, and finally Sect. 4 contains the concluding remarks of the study.

2   The Fuzzy Linear Programming Model

Firstly, we define the LP problem with fuzzy constraints (Zimmermann [15, 16]) as the following optimization problem

$$\begin{aligned} \operatorname*{Max}_{x} \; z &= c'x + c_0 \\ s.t. & \\ Ax &\lesssim B \\ x &\geqslant 0 \end{aligned} \tag{1}$$

where $x \in \mathbb{R}^n$, $c \in \mathbb{R}^n$, $c_0 \in \mathbb{R}$, $A \in \mathbb{R}^{n \times m}$. $B \in \mathcal{F}(x)$, $\mathcal{F}(x)$ is the set of all fuzzy sets. Every set B_i is a linear fuzzy set defined as:

$$B_i \triangleq \begin{cases} 1, & f(x) \leqslant \check{b}_i \\ \dfrac{\hat{b}_i - f(x)}{\hat{b}_i - \check{b}_i}, & \check{b}_i \leqslant f(x) \leqslant \hat{b}_i \\ 0, & f(x) \geqslant \hat{b}_i \end{cases} \tag{2}$$

where $\hat{b}_i, \check{b}_i \in \mathbb{R}$.

A comprehensive way to solve fuzzy constrained problems is by using the *Soft Constraints* model introduced by Zimmermann [15, 16]. The method solves this kind of problems through an α-cut approach, which consists on defining a set of solutions $Z(x^*)$ to find a joint-optimal $\alpha - cut$ for $Z(x^*)$ and $\tilde{b}$. The method is summarized next:

- Calculate an inferior bound called $\check{z} = \text{Max}_x\{z = c'x \mid Ax \leqslant \check{b}, x \geqslant 0\}$.
- Calculate a superior bound called $\hat{z} = \text{Max}_x\{z = c'x \mid Ax \leqslant \hat{b}, x \geqslant 0\}$.
- Define the fuzzy set $Z(x^*)$ with boundaries $\check{z}$ and $\hat{z}$ and linear membership function
$$\tilde{z}(x^*) = \frac{c'x^* - \check{z}}{\hat{z} - \check{z}}, \ \forall \ c'x^* \in [\check{z}, \hat{z}].$$
- Create an auxiliary variable α and solve the following LP model

$$\text{Max } \{\alpha\}$$
$$s.t.$$
$$c'x + c_0 - \alpha(\hat{z} - \check{z}) = \check{z} \tag{3}$$
$$Ax + \alpha(\hat{b} - \check{b}) \leqslant \hat{b}$$
$$x \geqslant 0$$

- Return α^*, x^*, b^*, z^*.

This method uses α as a decision variable that finds the Max intersection among all fuzzy constraints i.e. the maximum satisfaction degree among the goal and fuzzy constraints.

3 Concepts of Optimality Under Fuzzy Uncertainty

In this section, some aspects about the concept of an optimal solution under fuzzy constraints are addressed, starting from the meaning of feasible solutions, fuzzy optimal solution, and the idea of fuzzy global optimal solution.

The concept of a feasible solution in crisp LP is based on the idea of having a convex halfspace whose elements are feasible (or possible). In our case, this concept can be extended to an LP which has no crisp boundaries but fuzzy.

Now, the vector of boundaries $\hat{b}$ generates a halfspace namely $h(\cdot)$ which is the set of all values of x contained into the support of B_i, $x \in supp(B_i)$. This lead to the following definition:

Definition 1 A fuzzy constrained LP is feasible only if the polyhedron (or polytope) generated by $h(\cdot)$ is a non-trivial set, that is:

$$\mathcal{P} = \{x \mid h(\cdot) \leqslant \hat{b}\}, \tag{4}$$

where $\mathcal{P}$ is a non-trivial set of solutions of a crisp LP model.

Here, $\mathcal{P}$ is the convex set of solutions given $\hat{b}$, that is, the set of solutions constrained by $\hat{b}$. Therefore, any feasible solution contained into $\mathcal{P}$ has an associated **membership degree**, which leads us to define the following

Definition 2 Let $x' \in \mathcal{P}$ any feasible solution of $\sum a_{ij}x'_j = b'_i$ where $b'_i \in h(\cdot)$. Then, the linear combination $\sum a_{ij}x'_j$ belongs to B_i with a membership degree $\mu_{B_i}(x')$.

This means that every feasible solution x' can be projected over B_i reaching a membership degree $\mu_{B_i}(x')$, which basically is equivalent to say that in a fuzzy environment, every feasible solution x' has a membership degree $\mu_{B_i}(x')$. Figure 1 shows what a fuzzy feasible solution is.

Fuzzy optimal solution The concept of *optimal solution* of a fuzzy constrained LP is close to the optimality concept in LP. While in a crisp LP we have that an optimal solution is a vector x^* for which the function $z = c'x$ is certainly maximal, in a fuzzy constrained LP we have a set of optimal solutions (Z) which is a function of $^\alpha B$. This leads us to the following definition

Definition 3 A fuzzy optimal solution is defined as a vector x^* for which $\exists x$: $\max\{^\alpha Z = c'x \mid Ax \leqslant {}^\alpha B, x \geqslant 0, \alpha \in [0,1]\}$, so $^\alpha Z(x^*) \geqslant {}^\alpha Z(x) \, \forall x \in \mathbb{B}. \, \mathbb{B} \subseteq \mathbb{R}^+$ is the set of all feasible values of x, and $^\alpha Z(x^*)$ is the optimal objective value of $c'x^*$ given α.

Hence, a fuzzy optimal Fig. 3 solution is a vector x^* which fulfills all constraints and obtains a maximal value of the goal $z'^* = c'x^*$, at a membership degree α'^*. Now, as we are using $^\alpha B$ as a crisp value, then the value of each $^\alpha Z(x^*)$ is a *crisp global optimal* solution.

Every particular value of z' is comes from a crisp LP (see Fig. 2) model, so what we have is an optimal solution $z' = c'x^*$ given a particular value α, projected into $^\alpha B$. Note that between $\check{z}$ and $\hat{z}$ there is an infinite amount of possible optimal solutions (see Algorithm 2).

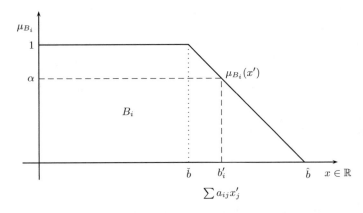

Fig. 1 Fuzzy feasible solution x' projected over B_i

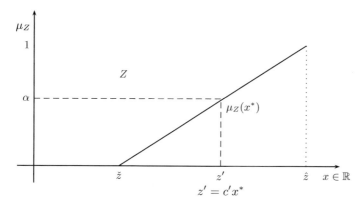

Fig. 2 Fuzzy optimal solution $z' = c'x^*$ projected over Z

Also note that as more α values are used, more x^* and $^{\alpha}Z(x^*)$ values exists. Indeed, there is an infinite amount of possible optimal solutions that can be computed, each one leading to a global optimal value given α.

Now, the concept of optimal solution in a fuzzy environment implies to obtain a fuzzy set which is a function of the parameters of the problem (and their membership functions). This leads to think about the concept what a *fuzzy global optimal* solution means.

3.1   Fuzzy Global Optimal Solution

The importance of having a *global optimal solution* in LP problems is large, since it is a key point for implementations and algorithms' design. In many applications, decision makers ask for a single (crisp) solution because they want to have a single operation point which returns the best possible results.

This is useful when a single solution *to be implemented* is requested, but in practice there is no reliability that optimal results can be applied. What sometimes happen is that the optimal solution cannot be implemented, so there is a need for having choices to be applied.

This makes sense to the fuzzy approach, since it obtains a set of optimal solutions that the analyst can use to compare the obtained results in practice to a set of possible choices. This allows to see how close (or far) are its results from the best possible solution.

In general, there is a need for having a relationship between **theoretical optimal** and what it is obtained in **practical applications**. Moreover, there is a need for clarify the sense of having an fuzzy optimal decision and what it means in practice. To do so, we have to take a look about concepts of global optimal solutions and its extension to a fuzzy environment.

The Bellman-Zadeh fuzzy decision making principle (see [1]) is a comprehensive way to solve fuzzy optimization problems. Its main idea is to solve a max − min decision making problem given known fuzzy sets which leads to find a single optimal intersection point between fuzzy constraints and the goal (see (3) in Algorithm 2, and (5)), and consequently a solution x^* which returns $^\alpha B$ and $^\alpha Z$.

Also, the Bellman-Zadeh fuzzy decision making principle (see [1]) (equivalent to the extension principle) allows us to find the membership degree of any solution of the problem. Bellman and Zadeh discussed in [1] a *hard* version of the extension principle which uses the max − min operators to solve unions and intersections, but they opened the door for using different *softer* versions for fuzzy decision making. This way, the Zimmermmann's approach computes the best α level from all combinations among the goal z^* and every binding constraint B_k:

$$\mu_{\tilde{z}}(z^*) = \sup_{z^*=\alpha^*} \min_k \{Z, B_1, \ldots, B_k, \ldots, B_K\} \tag{5}$$

Here, $Z = \mu_{\tilde{z}}(z^*)$ is the fuzzy set in which decision making is done, that is, the set from $z^* = \alpha^*$ is selected, and $B_k, k \in K$ is the k_{th} binding constraint. This means that α operates as an overall satisfaction degree of all fuzzy binding constraints, so α^* is the optimal defuzzification degree that reaches a crisp optimal solution of the problem.

This way, we can infer that a *fuzzy global solution* is then an *optimal solution given a fuzzy decision making criteria*. As usual, find a crisp point which fulfils both requirements could be expensive, so there is a need for using optimization methods able to handle fuzzy constraints while computing optimal solutions.

In the case of the Zimmermann's soft constraints method (see Sect. 2) its decision making optimization criterion is $\max_\alpha$, so the model shown in (1) operates as a crisp LP model that computes the maximum satisfaction degree among all constraints and the goal of the system.

Various authors have proven the existence of boundaries of optimal values of the goal function when solving fuzzy optimization problems, Ramík [13], Fiedler et al. [6], and Zimmermann [15, 16] have defined well known methods for finding the boundaries of the goal function ($\check{z}$ and $\hat{z}$ in this case).

This leads us to think in the following situation: *what is the usefulness of the solutions between $\check{z}$ and $\hat{z}$?*. This question leads us to think in all those points as alternatives to decision making in practical applications.

4 Ranking a Crisp Solution

The main idea addressed in this section is how to compare a solution achieved in practice to the optimal solution provided by the Zimmermann's method (see Sect. 2). To do so, we define what a feasible solution applied in practice is to later define a ranking method for comparing it to the optimal solution, and finally compute its membership degree to the set of optimal solutions.

4.1 Operation Points

From a practical point of view, there is no any certainty of reaching the optimal solution. When having a single optimal solution, the analyst should set the system in terms of that referring point in order to get its best performance.

If the analyst has choices or *operation points*, then decision making can be enriched because the analyst can use those points when setting the system, so basically if the system does not reach the expected results, the analyst can compare its current performance to a set of possible choices and see how good the performance of the system is. Then, an operation point is defined as follows:

Definition 4 (*Operation point*) An operation point is a set of observed values of b namely b' contained into the boundaries of B, $b' \in [\check{b}, \hat{b}]$ which leads to the optimal solution x^*, and z'.

An operation point is then an observed performance of the system which obeys to certain running conditions. What it is observed by a decision maker in real applications, is an operation point itself, so as many running conditions can occur as operation points can be compared by a decision maker.

To do so, we propose the following rank index for comparing an operation point (what was measured in real world) of the system against the optimal results after fuzzy decision making.

Definition 5 (*Ranking a solution*) Let b' a set of observed constraints $b' \in [\check{b}, \hat{b}]$, Z be the set of optimal solutions provided by any fuzzy decision making method, $z' \in Z$ be the optimal solution of the LP problem given b', α^* its optimal uncertainty degree, z^* the optimal solution of the fuzzy problem given α^*, and α' the membership degree of z' into Z. Then the relative degree of fulfilment ($Df_{z'}$) of z' compared to z^* is:

$$Df_{z'} = \frac{z' - z^*}{\hat{z} - \check{z}} \tag{6}$$

which is equivalent to

$$Df_{z'} = \alpha' - \alpha^* \tag{7}$$

It is clear that $Df_{z'} \in [-1, 1]$, so its interpretation is as follows: if $Df_{z'} > 0$ then the observed values of b' lead to improve the expected results; if $Df_{z'} < 0$ then observed values of b' did not reach the expected results, and if $Df_{z'} = 0$, then the values of b' have reached the expected results.

A comprehensive graph is provided in Fig. 3

Therefore, we can see that *every* operation point (b') observed in reality leads to x^*, z', α', having its own $Df_{z'}$. This allows us to compare the performance of the observed system and take actions to improve it. Moreover, $Df_{z'}$ allows us to compare different operation points at different stages of the system in order to make an appropriate decision.

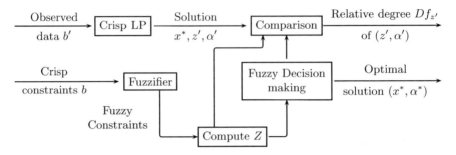

Fig. 3 Comparing observed values of b

On the other hand, it is necessary to see what is the membership degree reached by an operation point which comes from x^* regarding b'. As z' and α' comes from x^*, then we can use (5) to compute the membership degree of any operation point. To do so, we propose the following.

Proposition 1 *Let x^* be the optimal solution of an LP problem given b', $z' = c'x^*$ be its optimal value, $Z' = \mu_{Z'}(x^*)$ be the membership degree of x^* projected over Z, and $B'_i = \mu_{B'_i}(x^*)$ be the membership degree of x^* projected over B_i. Then, the membership degree of x^* given b' is:*

$$\mu_Z(z' : b') = \sup_{z'=c'x^*} \min_k \{Z', B'_1, \ldots, B'_k, \ldots, B'_K\} \tag{8}$$

where B'_k is the $k_{th} \in K$ binding constraint.

Note that (8) only involves binding constraints, since non-binding constraints does not provide an optimal extreme point. Also note that every operation point has a smaller membership degree over Z than the fuzzy global optimal solution provided by the Zimmermann's method (see (5)) since it provides the maximum satisfaction degree among all possible choices. This way, an operation point can overpass the value of z provided by the Zimmermann method, but it definitely has a smaller satisfaction degree of either the goal or a binding constraint.

Proposition 1 solves the intersection among constraints and the goal through the inf or min operator, and the union of all possible combinations of $z' = c'x^*$ through the sup or max operator. In the case of an optimal LP, we have only two cases: a single optimal solution x^* (the most probable case), and the multiple solutions case, so the sup operator makes sense only in the second case where multiple solutions should be compared. In the first case we only have a single solution, and there is no need for comparing it to other solutions.

4.2 Application Example

The application example has been taken from Klir and Folger [10], Example 15.8 at page 413. Assume that a company makes two products. Product P_1 has a $0.4 per unit profit and product P_2 has a $0.3 per unit profit. Product P_1 requires twice as many labor hours as each product P_2. The total available labor hours are at least 500 hours per day, and may be possible extended to 600 hours per day, due to some special arrangements for overtime work. The supply of material is at least sufficient for 400 units of both products P_1 and P_2, per day, but may be possible extended to 500 units per day according to previous experience. The problem is, how many units of products P_1 and P_2 should be made per day to maximize the total profit? The main problem can be expressed as follows

$$\text{Max}_x \ z = 0.4x_1 + 0.3x_2 \quad \text{(profit)}$$

$$\text{s.t.}$$
$$x_1 + x_2 \lesssim B_1 \quad \text{(material)}$$
$$2x_1 + x_2 \lesssim B_2 \quad \text{(labor hours)}$$
$$x_1, x_2 \geqslant 0$$

Then we have $\check{b} = [400; 500]$ and $\hat{b} = [500; 600]$. Using the Algorithm 2 (Zimmermann's method). the obtained results are $\check{z} = 130, \hat{z} = 160, z^* = 145, \alpha^* = 0.5, x_1^* = 100$ and $x_2^* = 350$.

Now suppose that the analyst did an experiment to try to set the system, and after all their attempts, the available labor hours and material were 530 and 415 units respectively, $b' = [415; 530]$. Then, the obtained results for this operation point were $z' = 136, \alpha' = 0.2, x_1^* = 115$ and $x_2^* = 300$. The relative degree of fulfilment of the current operation point is

$$Df_{z'} = \frac{136 - 145}{160 - 130} = -0.3 \tag{9}$$

The membership degree of this selection can be computed through (8):

$$Z' = 0.2, \ B_1' = 0.85, \ B_2' = 0.7$$
$$\mu_Z(z' : b') = \sup_{z'=136} \min_k \{0.2, 0.85, 0.7\} = 0.2$$

This means that this current operation point did not reach the expected results (since $Df_{z'} < 0$) and their membership degree is 0.2, so the analyst has to take actions to improve the system's performance.

Now suppose that the analyst has taken more actions to improve the availability of their resources. After some negotiations and improvements, it can increased the availablelabor hours and material to 560 and 465 units respectively, so $b' = [465; 560]$.

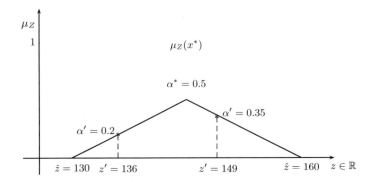

Fig. 4 Fuzzy optimal solution $z' = c'x^*$ projected over Z

Then, the obtained results for this *operation point* were $z' = 149$, $\alpha' = 0.6333$, $x_1^* = 95$ and $x_2^* = 370$. The relative degree of fulfilment of the current operation point is

$$Df_{z'} = \frac{149 - 145}{160 - 130} = 0.1333 \qquad (10)$$

The membership degree of this selection can be computed through (8):

$$Z' = 0.633, \quad B_1' = 0.35, \quad B_2' = 0.4$$
$$\mu_Z(z' : b') = \sup_{z'=149} \min_k\{0.633, 0.35, 0.4\} = 0.35$$

At this point, the current operation point of the system has overtaken its expected performance (since $Df_{z'} > 0$) with a membership degree of 0.35, so the analyst has taken actions which have improved the system's performance.

To illustrate how Definition 1 works, Fig. 4 displays the set Z and how operation points are located at.

Note that in the first operation point the goal is less than the second operation point, but both membership degrees are lesser than the optimal solution. In the first operation point there is a higher chance of getting all required resources to achieve $z' = 136$ while the second operation point has a smaller chance of getting all required resources to achieve $z' = 149$. As the goal increases, its chance of obtaining the required resources decreases, so the best possible solution sets an equilibrium between increasing the goal and the use of resources to increase it (which is reached by the Zimmermann's method).

Even when the second operation point has a $D_f > 0$ which means that z' overtakes the optimal solution, its membership degree regarding Z is lesser than α^* which means that it has a lesser possibility of occurrence. For the sake of understanding, the second operation point has a smaller possibility of occurrence than the first operation point which in practice means that more efforts are needed to obtain more resources to increase the goal, which is less possible. As the goal goes to $\hat{z} = 160$

then its possibility of occurrence decreases, and the system has to incur in higher efforts to increase z.

5 Concluding Remarks

A ranking measure to compare operation points to optimal solutions in FLP problems has been presented. We have applied it in two operation points over a theoretical application example, and analyzed their results.

A version of the extension principle to FLPs is proposed to compute the membership degree that an operation point has regarding the set of optimal solutions. We have applied its results to two cases, and we analyzed their results over Fig. 4.

We also clarify some concepts of fuzzy optimization that are applied in practice by providing a way to compare different solutions that can be reached in practice to the theoretical optimal solution. This helps the analyst to see how far/close is the current operation point of the system from its optimal solution.

Finally, we have explained how different operation points should be analyzed in order to make better decisions. The main idea is always obtain a higher goal, but in practice it implies to use more resources which is less possible, so the analyst can apply our results to establish an equilibrium between a higher goal using the less possible resources.

References

1. Bellman, R.E., Zadeh, L.A.: Decision-making in a fuzzy environment. Manag. Sci. **17**(1), 141–164 (1970)
2. Campos, L.: Fuzzy linear programming models to solve fuzzy matrix games. Fuzzy Sets Syst. **326**(1), 275–289 (1989)
3. Campos, L., Verdegay, J.: Linear programming problems and ranking of fuzzy numbers. Fuzzy Sets Syst. **32**(1), 1–11 (1989)
4. Černý, M., Hladík, M.: Optimization with uncertain, inexact or unstable data: linear programming and the interval approach. In: Němec, R., Zapletal, F. (eds.) In: Proceedings of the 10th International Conference on Strategic Management and its Support by Information Systems, Technical University of Ostrava, VŠB pp. 35–43 (2013)
5. Delgado, M., Verdegay, J.L., Vila, M.: A general model for fuzzy linear programming. Fuzzy Sets Syst. **29**(1), 21–29 (1989)
6. Fiedler, M., Nedoma, J., Ramík, J., Rohn, J., Zimmermann, K.: Linear Optimization Problems with Inexact Data. Springer (2006)
7. Friedman, M., Ming, M., Kandel, A.: Duality in fuzzy linear systems. Fuzzy Sets Syst. **109**, 55–58 (2000)
8. Hladík, M.: Weak and strong solvability of interval linear systems of equations and inequalities. Linear Algebr. Appl. **438**(11), 4156–4165 (2013)
9. Inuiguchi, M., Ramík, J.: Possibilistic linear programming: a brief review of fuzzy mathematical programming and a comparison with stochastic programming in portfolio selection problem. Fuzzy Sets Syst. **111**, 3–28 (2000)

10. Klir, G.J., Folger, T.A.: Fuzzy Sets, Uncertainty and Information. Prentice Hall (1992)
11. Mahdavi-Amiri, N., Nasseri, S.: Duality results and a dual simplex method for linear programming problems with trapezoidal fuzzy variables. Fuzzy Sets Syst. **158**(17), 1961–1978 (2007)
12. Ramík, J., Řimánek, K.: Inequality relation between fuzzy numbers and its use in fuzzy optimization. Fuzzy Sets Syst. **16**, 123–138 (1985)
13. Ramík, J.: Soft computing: overview and recent developments in fuzzy optimization. Technical Report, Institute for Research and Applications of Fuzzy Modeling Institute for Research and Applications of Fuzzy Modeling(2001)
14. Ramík, J.: Optimal solutions in optimization problem with objective function depending on fuzzy parameters. Fuzzy Sets Syst. **158**(17), 1873–1881 (2007)
15. Zimmermann, H.J.: Fuzzy programming and Linear Programming with several objective functions. Fuzzy Sets Syst. **1**(1), 45–55 (1978)
16. Zimmermann, H.J., Fullér, R.: Fuzzy Reasoning for solving fuzzy Mathematical Programming Problems. Fuzzy Sets Syst. **60**(1), 121–133 (1993)

On Modeling Multi-experts Multi-criteria Decision-Making Argumentation and Disagreement: Philosophical and Computational Approaches Reconsidered

Luciana Garbayo, Martine Ceberio, Stefano Bistarelli and Joel Henderson

1 Introduction

In an evidence-based paradigm aiming at technical decision-making, expert analysis provides necessary support for action and policy making. Such analyses include both knowledge and value considerations. Yet, unfortunately, a high level of expertise is not easily accessible in every needed circumstance, and, even when readily available, it may be notably tangled in expert disagreement among experts, generating even more complicated decision-making problems. Given its clear significance for professional problem-solving, both philosophers and computer scientists investigate and model expertise, expert disagreement and its entanglements, while studying expert decisions formally and by computational simulations. In this work, we focus primarily on collaboratively presenting some philosophical as well as computational aspects of argumentation and disagreement in the case of multi-experts multi-criteria decision-making (ME-MCDM) situations, aiming at contributing to the further goal of improving modeling expert decision analysis in an interdisciplinary endeavor.

Epistemology is the field of philosophy dedicated to the inquiry on the nature of knowledge, taking both propositional and non-propositional accounts thereof. Recently, a new sub-field of epistemology emerged, especially dedicated to the

L. Garbayo (✉) · M. Ceberio · J. Henderson
University of Texas at El Paso, El Paso, TX 79968, USA
e-mail: lsgarbayo@utep.edu

M. Ceberio
e-mail: mceberio@utep.edu

J. Henderson
e-mail: jahenderson@miners.utep.edu

S. Bistarelli
University of Perugia, 06123 Perugia, Italy
e-mail: bista@dmi.unipg.it

© Springer International Publishing AG 2018
M. Ceberio and V. Kreinovich (eds.), *Constraint Programming and Decision Making: Theory and Applications*, Studies in Systems, Decision and Control 100, DOI 10.1007/978-3-319-61753-4_10

study of disagreement, providing new clues to the understanding of the philosophical underpinnings thereof. This new sub-field is specially interested in the study of peer disagreement, in the narrow epistemic sense. Such precise contribution of philosophy fits quite well with the research aims of the constraint computing area, in the study of the challenges for modeling expert disagreement in the context of multi-criteria decision-making.

In this article, we specifically explore the epistemic modeling of argumentation among experts as epistemic peers, in the context of the work of Stefano Bistarelli and Martine Ceberio [1, 8], and Dung's argumentation theory [4]. In particular, we here consider the role of epistemic justification in expert disagreement on propositional contexts of knowledge, as well as in non-propositional contexts. We model the epistemology of disagreement of experts after Christensen's considerations on the role of formal constraints for rational belief, in reference to his excellent volume *Putting Logic in its Place* [2], and in reference to the debates on state of the art of the epistemology of disagreement (Epistemology of disagreement [3]), as well as our own interdisciplinary dialogue on the constraints to knowledge justification in expert disagreement [5].

2   Conceptualizing Disagreement Among Experts as Disagreement Among Epistemic Peers

The philosophical concept of disagreement in the domain of epistemology of disagreement is perhaps more precise and rare than in computer science, albeit both derive from ordinary daily experience of some sort of discord through rational discourse. In general philosophical terms, the ordinary experience of disagreement is mostly conceived first truth-functionally and in linguistic terms, broadly speaking. In this sense, the possibility of disagreement depends on someone sharing enough common understanding, so that one may disagree, on propositional matters. In other words, one should be able to understand semantics, syntactic, and pragmatic aspects of discourse of a speaker, to make sure that we understand what her point is, and yet possibly disagree with such and such position she puts forth, and/or propositional assertion. In this sense, a disagreement may be said to be properly legitimate, when those basic intelligibility conditions are met, or instead, verbal, when those conditions are not met, and we have a misunderstanding, rather than a disagreement.

In more strict epistemic terms, if such intelligibility conditions are taken to be met prima facie, legitimate disagreements are bound to occur as epistemic conflicts; see, e.g., [7]. Those happen when people are faced with conflicting beliefs that require the engagement in a belief revision process, in order to adjust one's confidence, so that one's belief is to be taken to be rational, in light of a certain particular disagreement [3]. In general, the degree of belief revision an epistemic agent under disagreement will engage depends on—as Christensen and Leakey contend—two dimensions of her appraisal of the epistemic credentials of her opponent (or opponents) on the matter under disagreement:

The first dimension of epistemic appraisal is related to the epistemic agent's evaluation of the level of familiarity of her opponent with evidence and arguments, in the case under dispute [3]. If there is asymmetry of familiarity, belief revision may (or may not) be required.

The second dimension is related to her epistemic appraisal of the opponent's competence at correctly evaluating both evidence and arguments [3]. Here cognitive and methodological aspects of knowing are considered. Under this framework, 'epistemic peers' are then said of those epistemic agents under disagreement who are appraised to be roughly equal along both dimensions [3].

In the same vein, we here suggest that experts under disagreement would then ideally be conceptualized as a special case of epistemic peers. Under this revised conceptualization, the second dimension of epistemic appraisal may be interpreted as being more fundamental than the first. For this reason, the expert should be taken to be first formally and methodologically apt to evaluate different scenarios. This dimension thus relates to general readiness on one's field of expertise. Disagreements of the sort of the first dimension—such as on familiarity or access to evidence and arguments—refer to the extension of one's expertise to cases, which may be theoretical, applied or practical.

In the next section, we will connect those two dimensions of epistemic appraisal of expertise under disagreement and consider the problem of rationality in expert belief-revision.

3 Expert Disagreement: Epistemic and Pragmatic Rationality

When experts legitimately disagree, they may do so on grounds of lack of completeness in knowing evidence and arguments. But also, while recognizing each other's cognitive competence, they may also recognize to be further applying different, alternative, or complementary methodologies and modeling techniques as cognitive strategies to one's field. In this case, experts might be using multi-criteria that may be incompatible as a whole, when contrasted; see, e.g., [9]. In this sense, experts under disagreement using different evaluative criteria would be in weak compliance with the second dimension of epistemic appraisal. For this reason, one may recognize cognitive competence, but not necessarily be ready to discuss the minutia of alternative expert criteria. See Table 1.

In the last case of weak compliance, there is understandably more uncertainty. In order to model a legitimate expert disagreement in the context of solving multi-criteria decision-making problems among experts, it is further useful to consider a philosophical distinction between epistemic and pragmatic rationality, for experts.

Table 1 Epistemic expert disagreement in two dimensions and Multi-criteria

First dimension of epistemic appraisal	Familiarity of the opponent with evidence and arguments
Second dimension of epistemic appraisal	Competence of the opponent at correctly evaluating evidence and arguments
Epistemic peers	If both appraisals result on roughly equal opponents
Experts under disagreement as epistemic peers	A strong compliance with second dimension, recognition of cognition and methodological expertise of opponent
Experts under disagreement as epistemic peers with different methodological approaches	A weak compliance with second dimension, where there is a shared knowledge of one's discipline, but the use of different expert criteria, generates further uncertainty

3.1 Epistemic Rationality: A More Subtle Focus of Disagreement on Epistemic Justification

When we consider the dimensions of epistemic appraisal to determine what disagreement is the case, we do so mostly under the light of our epistemic rationality. Here what is meant is to say that such revisions are descriptive in nature. They refer to the expert description of states of affairs of the world, according to the expert domain of knowledge in case, and to the matter of fact recognition of expert methodology. Here peer disagreements would more subtly appear in epistemic criteria for justification in argumentation and modeling concerns, regarding domain phenomena, and would mostly be settled on the grounds of assenting to graded beliefs, mostly based (Table 2) on probability of events or states of affairs established from scientific studies, rather then on simple binary beliefs. We dub such expert disagreement, disagreements on "epistemic justification" (see [5]).

3.2 Pragmatic Rationality

Let's imagine another relevant scenario for multi-criteria decision making among experts. If there is instead expert agreement at the epistemic level of description

Table 2 Epistemic expert disagreement in epistemic justification

Expert disagreement under epistemic rationality appraisal	Disagreement on the epistemic justification and epistemic criteria used in argumentation and modeling regarding domain phenomena

Table 3 Pragmatic expert disagreement

Pragmatic rationality apraisal	When applied uses are in discord and pragmatic justifications

of what is the case in the domain, in the three dimensions of appraisal mentioned, there may be an important disagreement on the applied uses of knowledge of graded beliefs, as suggested for purposes of decision-making. In this case, a new special type of expert disagreement may emerge, on the pragmatic use of such knowledge, with a concern with practical reason. This situation refers to concerns with pragmatic rationality, on value reasoning. Here the resulting model agreement may be seen as epistemic, but with a diverse practical result. Different sets of pragmatic justification may be salient (Table 3) embedded in description, according to what is deemed beneficial to decision-making. We dub such disagreement "pragmatic".

Thus, we here can have both epistemic and pragmatic types of disagreement, with fine-grain considerations on epistemic—evidence, argumentation, methodology issues—or pragmatic—disagreement on practical considerations.

3.3 Synchronic and Diachronic Rationality, Global and Local

Given this basic conceptualization of expert disagreement provided, we are ready to make further temporal and boundary considerations. It may be that experts disagree with themselves over time, and may be in contradiction, if considered contrasted in periods. In this case, we are modeling a dynamic picture of a disagreement with oneself and with others, possibly non-monotonic, describing change over time. This is the case of considering a synchronic and diachronic rationality for describing expert disagreements.

Another important concept is that of the boundaries of expertise. It is invariably the case that access to data or methodology is not absolute, thus, expertise is, indeed, always local, albeit possibly the most generalizable. The consideration of a completeness of data and methodology is an idealization, and much is uncertain. Thus, all experts are, in some sense, governed by a local (Table 4) rationality, and so are their disagreements, which may be restricted by parochialisms, and limitations of evidence.

Table 4 Expert disagreement (Types)

Epistemic	3 dimensions
Pragmatic	applied uses
Synchronic and diachronic	temporal considerations
Global and local	boundary considerations

4 Computational Modeling: Descriptive Constraints for Epistemic and Pragmatic Disagreements

We model both epistemic and pragmatic types of expert disagreement, with three main descriptive constraints: semantics, formal, and temporal.

- First, we distinguish semantically verbal from genuine type of disagreement, constraining the domain of debate. A verbal disagreement pertains to the misuse or misunderstanding of language use and terminology among experts, resulting on talking past each other. On the second case, experts understand each others special language use, and correctly disagree on the substance of what is been argued.
- Secondly, we use a global approach to formal constraints to rational belief, persuasive to compel rational belief revision in epistemic context, with deductive consistency and deductive closure.
- Thirdly, we consider consistency in synchronic descriptions and justifications of beliefs in time frames (simultaneous set of beliefs) and in rational belief revision over time in decision-making, related to diachronic description thereof (diachronic set of beliefs, consistency over time). As a complication, we consider the problem that one cannot do everything at once in epistemology [2], with some of the issues on parsing epistemic tasks.

Within the above-mentioned constraints, we model both epistemic and pragmatic expert disagreement with an argumentation framework based on Dungs semantics [1], represented with a graph approach, with weighted AF, for graded beliefs and a consideration of a trust score associated with fuzzy measures (Martines ref.). Such trust score takes in consideration the differences between epistemic and pragmatic contexts for establishing degrees of trust. In epistemic context, it aims at measuring expert trust in considering epistemic justification for rational belief against evidence available and argumentation. In pragmatic context, in order to evaluate pragmatic decision-making under multi-criteria decision-making criteria among experts, we discuss parameters of local rationality, as well as multi-criteria decision-making aggregation criteria, considering mono-dimensional utilities and the construction of a global utility function for each expert, and the aggregated results of their choices.

5 Preliminary Notions About Argumentation Frameworks and MEMCDM

The Dung's framework [4] allows to model argumentation through a graph of arguments (nodes) and attacks (edges). This framework was extended by Bistarelli et Al. [1] to elicit coalitions of arguments (e.g., through the notion of conflict-free sets or α-conflict-free sets).

In [6], we proposed to model MEMCDM using argumentation frameworks and to seek conflict-free or α-conflict-free sets as decision solutions. More specifically, let us look at a scenario in which experts independently assess given pieces of software, based on several given evaluation criteria. We illustrate our model on this problem:

5.1 Arguments

What information/arguments do we have to integrate into an Argumentation Network?

1. Expert i gives Item j a total quality $D_{i,j}$ (which, in the case of Software Quality Assessment SQA, can be Bad, Poor, Fair, Good, or Excellent):

$$\text{Argument} \quad (E_i, S_j, D_{i,j}).$$

2. Expert E_i judges that Software (or Item) S_j satisfies criterion m up to quality $D_{i,j,m}$:

$$\text{Argument} \quad (E_i, S_j, c_m, D_{i,j,m}).$$

5.2 Attacks

What are the attacks (edges of the network) between these arguments (nodes)?

1. Two experts disagree on their assessment of a given item:

$$\forall i, \ j, \ k, \ l, \ m \text{ such that } E_i \neq E_j \text{ and } D_k \neq D_l,$$

then there is an attack:

$$(E_i, S_k, D_l) \ \leftrightarrow \ (E_j, S_k, D_m).$$

2. For a given item, two experts disagree on their assessment of it w.r.t. a given criterion:

$$\forall i, \ j, \ k, \ l, \ m, \ n \text{ such that } E_i \neq E_j \text{ and } D_k \neq D_l,$$

then there is an attack:

$$(E_i, S_k, c_l, D_m) \ \leftrightarrow \ (E_j, S_k, c_l, D_n).$$

3. Every expert's decision is supported by his/her decisions on the criteria used to assess the item, but supports are not attacks: they are complementary. So support is expressed by an attack from every node that does not support.

6   How Epistemic and Pragmatic Disagreements Can Help MEMCDM

As of now, the model described in the previous section allows to describe decision processes as they relate to and are supported by assessment of intermediate criteria, and allows to model disagreement between experts, at the final decision level as well as at the intermediate criteria level.

This model is then processed so as to identify (α-) conflict-free sets: these sets should contain arguments that are not attacked (unlikely) or the least attacked (i.e., the least controversial among experts, which corresponds to alpha-conflict-free sets for the lowest possible α). However, disagreement remains in these cases: only a solution (even the least controversial) has been reached.

The purpose of this work is to expand on such modeling to further help de- tangling disagreements. The assumption is that disagreement is made of either or both epistemic and pragmatic disagreements. As currently modeled, decision processes do not differentiate these two decision levels, so when identifying disagreement (as modeled in Sect. 4), we do not have access to fine-grained information about why a disagreement occurs and how to diffuse/understand it. By acknowledging that disagreement can be epistemic or pragmatic, and by further modeling an argumentation framework for each of these dimensions (or levels of decisions), we can exploit argumentation frameworks (as proposed by Bistarelli et al. [1]) further. We propose to model decision process as two argumentation graphs: one that models the epistemic reasoning of experts, and the other one to model their pragmatic reasoning.

For instance, we could identify that there is disagreement on the epistemic level but not on the pragmatic level, or vice versa, or both. No disagreement (resp. low disagreement) would result ([7]) in an edge-less graph (resp. a solution that encompasses all nodes of the graph as the α-conflict-free set).

Having knowledge of the levels at which disagreement occurs could inform about how to resolve conflicts. For instance, if experts agree on the pragmatic level but not on the epistemic level, that might mean that they do not agree on the metrics of the problem but do agree on the general solution. Such a finding might call for a revision of the problem description: maybe the original decision criteria were not appropriate.

On the other hand ([9]) showing agreement at the epistemic level but not at the pragmatic level might be typical of experts who have different goals (different values).

7   What's Next?

In this work, focusing on decision-making situations involving multiple experts, and with the aim to identify and understand disagreement, we proposed a new modeling approach that relies on Dung's argumentation framework extended by Bistarelli. Our approach acknowledges that disagreement can be at two different levels: epistemic and pragmatic, and makes use of argumentation frameworks to identify disagreement configurations (epistemic and pragmatic, epistemic only, pragmatic only).

The next step to this work will consist in exploring ways in which disagreement identification can be used, not to diffuse disagreement (since, when data is available, experts are no longer around the table) but to elicit an automated decision-making process, via the addition of expertise levels and the notion of trust of individual decisions.

Acknowledgements This work was partially supported by the National Science Foundation, NSF CCF grant 0953339 and the American Association for the Advancement of Science, AAAS MIRC (agreement date 112612).

References

1. Bistarelli, S., Santini, F.: A common computational framework for semiring-based argumentation systems. Font. Artif. Intell. Appl.: ECAI 2010 **215**(131-136) (2010)
2. Christensen, D.: Putting Logic in its Place: Formal Constraints to Rational Belief. Oxford University Press (2007)
3. Christensen, D., Lackey, J. (eds.): The Epistemology Of Disagreement: New essays. Oxford University Press (2013)
4. Dung, P.M.: On the acceptability of arguments and its fundamental role in nonmono-tonic reasoning, logic programming and n-person games. Artif. Intell. **77**(2), 321–357 (1995)
5. Garbayo, L.: Epistemic considerations on expert disagreement, normative justification and inconsistency regarding multi-criteria decision-making. In Ceberio, M., Krenovich, V.: (eds.), Constraint Programming and Decision MakingStudies in Computational Intelligence Vol. **539**, pp. 35-45(2010)
6. Henderson, J., Bistarelli, S., Ceberio, M.: Multi-Experts Multi-Criteria Decision Making. Presented at the international conference on numerical computations: theory and algorithms, NUMTA 2013
7. Rami, B.: An overview of methods to elicit and model expert clinical judgment and decision making. Soc. Serv. Rev. **66**(4), 596–616 (1992)
8. Wang, X., Garcia Contreras, A. F., Ceberio, M., Del Hoyo, C., Gutierrez, L. C., Virani, S.: Interval-based algorithms to extract fuzzy measures for software quality assessment. In: North American Fuzzy Information Processing Society (NAFIPS'2012), Berkeley, CA, August 2012
9. Yager, R.: Non-Numeric Multi-Criteria Multi-Person Decision-Making. Group Decis. Negot. **2**, 81–93 (1993)

Positive Semidefiniteness and Positive Definiteness of a Linear Parametric Interval Matrix

Milan Hladík

1 Introduction

A commonly used deterministic approach to global optimization [3, 5, 7, 8, 15, 27] is based on exhaustive splitting of the search space into smaller parts (usually boxes) and applying various interval techniques to remove boxes that provably do not contain any global minimizer, to compute rigorous lower and upper bounds on the optimal value, and to prove optimality of some point within a box, among others.

An important step in this approach is convexity testing on a box. If the objective function is identified as convex on the box, any local minimum is also global, and the search within the box becomes easier. Similarly, if the function is convex nowhere on the box and the box lies inside the feasible set, then the box can be removed as it contains no local, and hence also no global, minimum. Convexity also plays an important role in the global optimization αBB method [3–5, 12, 39], which is based in constructing a convex underestimator of the objective function by appending an additional convex quadratic term.

Convexity of the objective function on a box is usually studied via an interval matrix enclosing all Hessian matrices of the function on the box. Since convexity of a function corresponds to positive (semi-)definiteness of its Hessian matrix, we face the problem of checking positive (semi-)definiteness of an interval matrix.

Let us introduce some notation now. We use $\operatorname{diag}(z)$ for the diagonal matrix with entries $z_1, \ldots, z_n$, and $\operatorname{sgn}(r)$ for the sign of r ($\operatorname{sgn}(r) = 1$ if $r \geq 0$ and $\operatorname{sgn}(r) = -1$ otherwise). For vectors, the sign is applied entrywise.

An interval matrix A is defined as

$$A := [\underline{A}, \overline{A}] = \{A \in \mathbb{R}^{m \times n}; \ \underline{A} \leq A \leq \overline{A}\},$$

M. Hladík (✉)
Faculty of Mathematics and Physics, Department of Applied Mathematics,
Charles University, Malostranské Nám. 25, 118 00 Prague, Czech Republic
e-mail: hladik@kam.mff.cuni.cz
URL:http://kam.mff.cuni.cz/~hladik/

© Springer International Publishing AG 2018
M. Ceberio and V. Kreinovich (eds.), *Constraint Programming and Decision Making: Theory and Applications*, Studies in Systems, Decision and Control 100,
DOI 10.1007/978-3-319-61753-4_11

where $\underline{A}, \overline{A} \in \mathbb{R}^{m \times n}$, $\underline{A} \leq \overline{A}$, are given, and the inequality is understood entrywise. *The midpoint* and *the radius* of A are defined respectively as

$$A^c := \frac{1}{2}(\underline{A} + \overline{A}), \quad A^\Delta := \frac{1}{2}(\overline{A} - \underline{A}).$$

The set of all m-by-n interval matrices is denoted by $\mathbb{IR}^{m \times n}$. Supposing that both A^c and A^Δ are symmetric, the symmetric counterpart to A is

$$A^S := \{A \in A; \ A = A^T\}.$$

A symmetric interval matrix $A^S \in \mathbb{IR}^{n \times n}$ is *strongly positive definite (positive semidefinite)* if A is positive definite (positive semidefinite) for each $A \in A^S$. Next, A^S is *weakly positive definite (positive semidefinite)* if A is positive definite (positive semidefinite) for some $A \in A^S$. Eventually, $A \in \mathbb{IR}^{n \times n}$ is *regular* if every $A \in A$ is nonsingular.

The classical results characterizing strong positive semidefiniteness and positive definiteness are stated below; see Rohn [33, 35, 36], and Białas and Garloff [1]. Suppose that $A \in \mathbb{IR}^{n \times n}$ is given with A^c and A^Δ symmetric.

Theorem 1 *The following are equivalent:*

(1) A^S *is positive semidefinite,*
(2) $A^c - \mathrm{diag}(z)A^\Delta\mathrm{diag}(z)$ *is positive semidefinite for each* $z \in \{\pm 1\}^n$,
(3) $x^T A^c x - |x|^T A^\Delta |x| \geq 0$ *for each* $x \in \mathbb{R}^n$.

Theorem 2 *The following are equivalent:*

(1) A^S *is positive definite,*
(2) $A^c - \mathrm{diag}(z)A^\Delta\mathrm{diag}(z)$ *is positive definite for each* $z \in \{\pm 1\}^n$,
(3) $x^T A^c x - |x|^T A^\Delta |x| > 0$ *for each* $0 \neq x \in \mathbb{R}^n$,
(4) A^c *is positive definite and* A *is regular.*

Checking strong positive (semi-)definiteness is known to be a co-NP-hard problem (Kreinovich et al. [19]). On the other hand, checking whether there is a positive semidefinite matrix in A^S is a polynomial time problem; see Jaulin and Henrion [14].

There are other related results on positive definiteness of interval matrices. For instance, Liu [21] presents a sufficient condition and applies it to stability issues, Kolev [17] presents a method to determine a positive definite margin of an interval matrix, and Shao and Hou [38] propose a necessary and sufficient criterion for a larger class of complex Hermitian interval matrices.

Positive (semi-)definiteness closely relates to matrix eigenvalues. A real symmetric matrix A is positive (semi-)definite if and only if all its eigenvalues are positive (nonnegative). This relation indicates that positive (semi-)definiteness can be investigated from the perspective of eigenvalues of interval matrices. Such eigenvalues were studied, e.g., in [11, 13, 16, 18, 20, 22, 25], and some of those results could possibly be used to check for positive (semi-)definiteness; a simple sufficient condition for strong positive definiteness appeared already in Rohn [33, 35, 36]. This paper, however, is focused in other direction. We generalize some of the classical results to interval matrices affected by linear dependencies between the matrix entries.

2 Linear Parametric Matrices: Positive Semidefiniteness

The standard notion of an interval matrix assumes that all matrix entries vary within the given intervals independently of other entries. This assumption is rarely satisfied in practice. To approach more closely to practical use and to model possible dependencies, consider a more general concept of a linear parametric matrix

$$A(p) = \sum_{k=1}^{K} A^{(k)} p_k,$$

where $A^{(1)}, \ldots, A^{(K)} \in \mathbb{R}^{n \times n}$ are fixed symmetric matrices and $p_1, \ldots, p_K$ are parameters varying respectively in $p_1, \ldots, p_K \in \mathbb{IR}$.
   Strong and weak positive definiteness extends to parametric matrices naturally as follows.

Definition 1 A parametric matrix $A(p)$, $p \in p$, is *strongly positive definite (positive semidefinite)* if $A(p)$ is positive definite (positive semidefinite) for each $p \in p$. It is *weakly positive definite (positive semidefinite)* if $A(p)$ is positive definite (positive semidefinite) for at least one $p \in p$.

   Linear parametric form generalizes the standard interval matrix. Evaluation $A(p) = \sum_{k=1}^{K} A^{(k)} p_k$ by interval arithmetic encloses the set of matrices $A(p)$, $p \in p$, in an interval matrix and reduces the problem to the standard non-parametric one. This "relaxation" of parametric structure, however, overestimates the true set and may lead to loss of positive (semi-)definiteness.

Example 1 Let

$$A(p) = \begin{pmatrix} 1 & 1 \\ 1 & 1 \end{pmatrix} p, \quad p \in p = [0, 1].$$

This parametric matrix is strongly positive semidefinite, but its relaxation

$$A(p) = \begin{pmatrix} [0, 1] & [0, 1] \\ [0, 1] & [0, 1] \end{pmatrix}$$

is not, as it contains, e.g., the indefinite matrix

$$\begin{pmatrix} 0 & 1 \\ 1 & 0 \end{pmatrix}.$$

□

Linear parametric forms are also used to model linear dependencies between parameters in interval linear equation solving [10, 30, 31, 41]. Linear dependencies cause not only the problem to be more difficult from the computational viewpoint, but it is also hard to describe the corresponding solution set; see Mayer [23].

2.1 Strong Positive Semidefiniteness

Surprisingly, characterization of strong positive semidefiniteness from Theorem 1 can be extended to parametric matrices quite straightforwardly.

Theorem 3 *The following are equivalent:*

(1) $A(p)$ is positive semidefinite for each $p \in \boldsymbol{p}$,
(2) $A(p)$ is positive semidefinite for each p such that $p_k \in \{\underline{p}_k, \overline{p}_k\}\ \forall k$,
(3) $x^T A(p^c) x - \sum_{k=1}^{K} |x^T A^{(k)} x| \cdot p_k^\Delta \geq 0$ for each $x \in \mathbb{R}^n$.

Proof "(1) $\Rightarrow$ (2)"
Obvious. "(2) $\Rightarrow$ (3)" Let $0 \neq x \in \mathbb{R}^n$. Define $s_k := \mathrm{sgn}(x^T A^{(k)} x)$ and $p_k := p_k^c - s_k p_k^\Delta \in \{\underline{p}_k, \overline{p}_k\}$, $k = 1, \ldots, K$. Now, by positive semidefiniteness of $A(p)$, we have

$$x^T A(p^c) x - \sum_{k=1}^{K} |x^T A^{(k)} x| \cdot p_k^\Delta = x^T A(p^c) x - \sum_{k=1}^{K} x^T A^{(k)} x s_k p_k^\Delta$$

$$= \sum_{k=1}^{K} x^T A^{(k)} x p_k = x^T A(p) x \geq 0.$$

"(3) $\Rightarrow$ (1)" Let $p \in \boldsymbol{p}$ and $x \in \mathbb{R}^n$. Now,

$$x^T A(p) x = \sum_{k=1}^{K} x^T A^{(k)} x p_k = x^T A(p^c) x + \sum_{k=1}^{K} x^T A^{(k)} x (p_k - p_k^c)$$

$$\geq x^T A(p^c) x - \sum_{k=1}^{K} |x^T A^{(k)} x| \cdot |p_k - p_k^c|$$

$$\geq x^T A(p^c) x - \sum_{k=1}^{K} |x^T A^{(k)} x| \cdot p_k^\Delta \geq 0.$$

□

This result shows that strong positive semidefiniteness can be verified by checking positive semidefiniteness of 2^K real matrices. This enables us to effectively check strong positive semidefiniteness of large matrices provided the number of parameters is small. Moreover, as stated below, the number 2^K can be further reduced in some cases.

Theorem 4 *(1) If $A^{(i)}$ is positive semidefinite for some i, then we can fix $p_i := \underline{p}_i$ for checking strong positive semidefiniteness.*
(2) If $A^{(i)}$ is negative semidefinite for some i, then we can fix $p_i := \overline{p}_i$ for checking strong positive semidefiniteness.

Proof (1) Let $p \in \boldsymbol{p}$. We use the fact that positive semidefiniteness is closed under addition and nonnegative multiples. Thus, $A^{(i)}(p_i - \underline{p}_i)$ is positive definite. If

$$\sum_{k \neq i} A^{(k)} p_k + A^{(i)} \underline{p}_i$$

is positive semidefinite for some $p_k \in \boldsymbol{p}_k$, $k \neq i$, then

$$\sum_{k \neq i} A^{(k)} p_k + A^{(i)} \underline{p}_i + A^{(i)}(p_i - \underline{p}_i) = A(p)$$

is positive semidefinite, too.
(2) Analogously.   □

As long as K is too large to apply Theorem 3, and Theorem 4 fails to reduce the number of real matrices to be processed, the following sufficient condition may be useful.

Theorem 5 *For each $k = 1, \ldots, K$, let $A^{(k)} = A_1^{(k)} - A_2^{(k)}$, where both $A_1^{(k)}, A_2^{(k)}$ are positive semidefinite. Then $A(p)$, $p \in \boldsymbol{p}$, is strongly positive semidefinite if*

$$\sum_{k=1}^{K} \left(A_1^{(k)} \underline{p}_k - A_2^{(k)} \overline{p}_k \right)$$

is positive semidefinite.

Proof Let $p \in \boldsymbol{p}$. By closedness of positive semidefiniteness under addition and nonnegative multiples, we have that

$$A(p) = \sum_{k=1}^{K} \left(A_1^{(k)} p_k - A_2^{(k)} p_k \right)$$

$$= \sum_{k=1}^{K} \left(A_1^{(k)} \underline{p}_k - A_2^{(k)} \overline{p}_k \right) + \sum_{k=1}^{K} \left(A_1^{(k)}(p_k - \underline{p}_k) + A_2^{(k)}(\overline{p}_k - p_k) \right)$$

is positive semidefinite, too.   □

A splitting of $A^{(k)}$ into a difference between two positive semidefinite matrices can be carried out as follows. Let $A^{(k)} = Q \Lambda Q^T$ be a spectral decomposition of $A^{(k)}$. Let Λ^+ be the diagonal matrix the entries of which are the positive parts of Λ, and similarly Λ^- has the negative parts on the diagonal. Then $A^{(k)} = Q \Lambda Q^T = Q \Lambda^+ Q^T - Q \Lambda^- Q^T$ and both $Q \Lambda^+ Q^T$, $Q \Lambda^- Q^T$ are positive semidefinite.

2.2 Weak Positive Semidefiniteness

Concerning weak positive semidefiniteness, the problem is still solvable in polynomial time by utilizing a suitable semidefinite program [6, 26, 40]. Let $M(p)$ be the block diagonal matrix with blocks

$$A(p), \ p_1 - \underline{p}_1, \ \ldots, \ p_K - \underline{p}_K, \ \overline{p}_1 - p_1, \ \ldots, \ \overline{p}_K - p_K.$$

All entries of $M(p)$ depend affinely on variables p. Positive definiteness of $M(p)$ is equivalent to positive definiteness of $A(p)$ and feasibility of variables $p \in \boldsymbol{p}$. Therefore, by solving this semidefinite program we check whether $A(p)$, $p \in \boldsymbol{p}$, is weakly positive semidefinite.

Anyway, a cheap necessary condition may be useful, e.g., for nonconvexity testing in global optimization [7].

Theorem 6 *For each $k = 1, \ldots, K$, let $A^{(k)} = A_1^{(k)} - A_2^{(k)}$, where both $A_1^{(k)}$, $A_2^{(k)}$ are positive semidefinite. If $A(p)$, $p \in \boldsymbol{p}$, is weakly positive semidefinite, then*

$$\sum_{k=1}^{K} \left(A_1^{(k)} \overline{p}_k - A_2^{(k)} \underline{p}_k \right)$$

is positive semidefinite.

Proof Let $p \in \boldsymbol{p}$ such that $A(p)$ is positive semidefinite. By closedness of positive semidefiniteness under addition and nonnegative multiples, we have that

$$A(p) + \sum_{k=1}^{K} \left(A_1^{(k)} (\overline{p}_k - p_k) + A_2^{(k)} (p_k - \underline{p}_k) \right) = \sum_{k=1}^{K} \left(A_1^{(k)} \overline{p}_k - A_2^{(k)} \underline{p}_k \right)$$

is positive semidefinite, too.   □

In view of Theorem 4, it is easy to see that the conditions from Theorems 5 and 6 are necessary and sufficient provided for each $k = 1, \ldots, K$, the matrix $A^{(k)}$ is either positive or negative definite.

3 Linear Parametric Matrices: Positive Definiteness

In a similar fashion as in Sect. 2, we can characterize positive definiteness of parametric matrices.

Theorem 7 *The following are equivalent:*

(1) $A(p)$, $p \in \boldsymbol{p}$, is strongly positive definite,

(2) $A(p)$ is positive definite for each p such that $p_k \in \{\underline{p}_k, \overline{p}_k\}$ $\forall k$,

(3) $x^T A(p^c) x - \sum_{k=1}^{K} |x^T A^{(k)} x| \cdot p_k^\Delta > 0$ for each $0 \neq x \in \mathbb{R}^n$.

Proof Analogous to Theorem 3. □

Theorem 8 *(1) If $A^{(i)}$ is positive semidefinite for some i, then we can fix $p_i := \underline{p}_i$ for checking strong positive definiteness.*

(2) If $A^{(i)}$ is negative semidefinite for some i, then we can fix $p_i := \overline{p}_i$ for checking strong positive definiteness.

Proof Analogous to Theorem 4. □

Theorem 9 *For each $k = 1, \ldots, K$, let $A^{(k)} = A_1^{(k)} - A_2^{(k)}$, where both $A_1^{(k)}, A_2^{(k)}$ are positive semidefinite. Then $A(p)$, $p \in \boldsymbol{p}$, is strongly positive definite if*

$$\sum_{k=1}^{K} \left(A_1^{(k)} \underline{p}_k - A_2^{(k)} \overline{p}_k \right)$$

is positive definite.

Proof Analogous to Theorem 5. □

A parametric matrix $A(p)$, $p \in \boldsymbol{p}$, is called *regular* if $A(p)$ is nonsingular for each $p \in \boldsymbol{p}$. Regularity of parametric matrices was investigated by Popova [29], for instance. We have the following relation to regularity, extending item (4) of Theorem 2.

Theorem 10 *The parametric matrix $A(p)$, $p \in \boldsymbol{p}$, is strongly positive definite if and only if the following two properties hold:*

(1) $A(p)$ is positive definite for an arbitrarily chosen $p \in \boldsymbol{p}$,

(2) $A(p)$, $p \in \boldsymbol{p}$, is regular.

Proof "$\Rightarrow$" Obvious as each positive definite matrix is nonsingular.

"$\Leftarrow$" Let $A(p^1)$ be positive definite for $p^1 \in \boldsymbol{p}$ and suppose to the contrary that $A(p^2)$ is not positive definite for $p^2 \in \boldsymbol{p}$. Hence $A(p^1)$ has positive eigenvalues, and $A(p^2)$ has at least one non-positive eigenvalue. Due to continuity of eigenvalues [24] and compactness of $\boldsymbol{p}$, there is $p^0 \in \boldsymbol{p}$ such that $A(p^0)$ is singular. A contradiction. □

Now, we have two sufficient conditions for checking strong positive definiteness. The first one is stated in Theorem 9, and the second one utilizes regularity according to Theorem 10. By Poljak and Rohn [28] (see also [2, 19]), checking regularity of an interval matrix is a co-NP-hard problem, but there are some polynomially verifiable sufficient conditions; see Rex and Rohn [32]. The commonly used one, the Beeck criterion, checks whether $\rho(M^\Delta) < 1$, where

$$M := \sum_{k=1}^{K} (C A^{(k)}) p_k,$$

and $C = A(p^c)^{-1}$ is the preconditioner.

We show by examples that no one condition for checking strong positive definiteness is stronger than the other one, where the Beeck criterion is utilized for regularity checking. Notice that the values in the matrices below are displayed to a precision of four digits, however, the real computation was done rigorously in Matlab using the interval library Intlab by Rump [37] and the verification software package Versoft by Rohn [34].

Example 2 Let

$$A(p) = \begin{pmatrix} 1.5 & 0 \\ 0 & 1.1 \end{pmatrix} p_1 + \begin{pmatrix} -1 & 1 \\ 1 & 1 \end{pmatrix} p_2, \quad p \in \boldsymbol{p} = (1, [0, 1]).$$

This parametric matrix is strongly positive definite.

Let us check the sufficient condition by Theorem 9. The matrix $A^{(1)}$ is positive definite, so we split only

$$A^{(2)} = A_1^{(2)} - A_2^{(2)} = \begin{pmatrix} 0.2071 & 0.5 \\ 0.5 & 1.2071 \end{pmatrix} - \begin{pmatrix} 1.2071 & -0.5 \\ -0.5 & 0.2071 \end{pmatrix}.$$

Now,

$$A^{(1)} \cdot 1 + A_1^{(2)} \cdot 0 - A_2^{(2)} \cdot 1 = \begin{pmatrix} 0.2929 & 0.5 \\ 0.5 & 0.8929 \end{pmatrix}$$

is positive definite, proving strong positive definiteness of $A(p)$, $p \in \boldsymbol{p}$.

In comparison, the Beeck sufficient regularity condition fails to prove regularity. Using the preconditioner

$$C := A(p^c)^{-1} = \begin{pmatrix} 1 & 0.5 \\ 0.5 & 1.6 \end{pmatrix}^{-1},$$

the relaxation leads to an interval matrix

$$M = \sum_{k=1}^{K} (C A^{(k)}) p_k = \begin{pmatrix} [0.2222, 1.7778] & [-0.4075, 0.4075] \\ [-0.5556, 0.5556] & [0.8148, 1.1852] \end{pmatrix},$$

which is not confirmed to be regular by using the Beeck condition as $\rho(M^\Delta) = 1.0419 \not< 1$. □

Example 3 Let

$$A(p) = \begin{pmatrix} 3.3 & 0.25 \\ 0.25 & 3.3 \end{pmatrix} p_1 + \begin{pmatrix} 1 & 2 \\ 2 & 0 \end{pmatrix} p_2 + \begin{pmatrix} 0 & 2 \\ 2 & 1 \end{pmatrix} p_3, \quad p \in \mathbf{p} = (1, [0, 1], [0, 1]).$$

In this example, Theorem 9 fails to prove positive definiteness. In contrast, with the preconditioner

$$C := A(p^c)^{-1} = \begin{pmatrix} 3.8 & 2.25 \\ 2.25 & 3.8 \end{pmatrix}^{-1},$$

and the relaxation matrix

$$M = \sum_{k=1}^{K} (CA^{(k)}) p_k = \begin{pmatrix} [0.7227, 1.2773] & [-0.6905, 0.6905] \\ [-0.6905, 0.6905] & [0.7227, 1.2773] \end{pmatrix},$$

the Beeck condition proves positive definiteness by showing $\rho(M^\Delta) = 0.9678 < 1$.

Theorem 6 is modified to necessary condition for weak positive definiteness as follows.

Theorem 11 *For each $k = 1, \ldots, K$, let $A^{(k)} = A_1^{(k)} - A_2^{(k)}$, where both $A_1^{(k)}, A_2^{(k)}$ are positive semidefinite. If $A(p)$, $p \in \mathbf{p}$, is weakly positive definite, then*

$$\sum_{k=1}^{K} \left(A_1^{(k)} \overline{p}_k - A_2^{(k)} \underline{p}_k \right)$$

is positive definite.

Proof Analogous to Theorem 6. □

4 Example

Consider a class of functions

$$f(x) = \sum_{\ell=1}^{L} c_\ell x_{i_\ell} x_{j_\ell} x_{k_\ell},$$

where $i_\ell, j_\ell, k_\ell \in \{0, \ldots, n\}$ are not necessarily mutually different, and $x_0 = 1$. For such functions, their Hessian matrix has directly a linear parametric form without using any kind of linearization. It is easy to see that each entry of the Hessian of

$f(x)$ is a linear function with respect to $x \in \mathbb{R}^n$. Thus the variables x play the role of the parameters p, and their domain $\boldsymbol{x}$ works as $\boldsymbol{p}$.

Example 4 Let

$$f(x, y, z) = x^3 + 2x^2 y - xyz + 3yz^2 + 5y^2,$$

and we want to check its convexity on $x \in \boldsymbol{x} = [2, 3]$, $y \in \boldsymbol{y} = [1, 2]$ and $z \in \boldsymbol{z} = [0, 1]$. The Hessian of f reads

$$\nabla^2 f(x, y, z) = \begin{pmatrix} 6x + 4y & 4x - z & -y \\ 4x - z & 10 & -x + 6z \\ -y & -x + 6z & 6y \end{pmatrix}$$

The direct evaluation the Hessian matrix by interval arithmetic results in an enclose by the interval matrix

$$\nabla^2 f(\boldsymbol{x}, \boldsymbol{y}, \boldsymbol{z}) \subseteq \begin{pmatrix} [16, 26] & [7, 12] & -[1, 2] \\ [7, 12] & 10 & [-3, 4] \\ -[1, 2] & [-3, 4] & [6, 12] \end{pmatrix}$$

This interval matrix is not strongly positive semidefinite since the smallest eigenvalue, computed by the exponential formula by Hertz [9], is -2.8950. Nevertheless, Theorem 9 proves $\nabla^2 f(x, y, z)$ to be positive definite by utilizing the parametric form

$$\nabla^2 f(x, y, z) = \begin{pmatrix} 6 & 4 & 0 \\ 4 & 0 & -1 \\ 0 & -1 & 0 \end{pmatrix} x + \begin{pmatrix} 4 & 0 & -1 \\ 0 & 0 & 0 \\ -1 & 0 & 6 \end{pmatrix} y + \begin{pmatrix} 0 & -1 & 0 \\ -1 & 0 & 6 \\ 0 & 6 & 0 \end{pmatrix} z + \begin{pmatrix} 0 & 0 & 0 \\ 0 & 10 & 0 \\ 0 & 0 & 0 \end{pmatrix}.$$

Thus, we can conclude that f is convex on the interval domain.

Acknowledgements The author was supported by the Czech Science Foundation Grant P402-13-10660S.

References

1. Białas, S., Garloff, J.: Intervals of P-matrices and related matrices. Linear Algebra Appl. **58**, 33–41 (1984)
2. Fiedler, M., Nedoma, J., Ramík, J., Rohn, J., Zimmermann, K.: Linear Optimization Problems with Inexact Data. Springer, New York (2006)
3. Floudas, C.A.: Deterministic Global Optimization. Theory, Methods and Application. In: Nonconvex Optimization and its Applications, vol. 37. Kluwer, Dordrecht (2000)
4. Floudas, C.A., Gounaris, C.E.: A review of recent advances in global optimization. J. Glob. Optim. **45**(1), 3–38 (2009)

5. Floudas, C.A., Pardalos, P.M. (eds.): Encyclopedia of Optimization, 2nd edn. Springer, New York (2009)
6. Gärtner, B., Matoušek, J.: Approximation Algorithms and Semidefinite Programming. Springer, Berlin Heidelberg (2012)
7. Hansen, E.R., Walster, G.W.: Global Optimization Using Interval Analysis, 2nd edn. Marcel Dekker, New York (2004)
8. Hendrix, E.M.T., Gazdag-Tóth, B.: Introduction to nonlinear and global optimization. In: Optimization and Its Applications, vol. 37. Springer, New York (2010)
9. Hertz, D.: The extreme eigenvalues and stability of real symmetric interval matrices. IEEE Trans. Autom. Control **37**(4), 532–535 (1992)
10. Hladík, M.: Enclosures for the solution set of parametric interval linear systems. Int. J. Appl. Math. Comput. Sci. **22**(3), 561–574 (2012)
11. Hladík, M.: Bounds on eigenvalues of real and complex interval matrices. Appl. Math. Comput. **219**(10), 5584–5591 (2013)
12. Hladík, M.: On the efficient Gerschgorin inclusion usage in the global optimization αBB method. J. Glob. Optim. **61**(2), 235–253 (2015)
13. Hladík, M., Daney, D., Tsigaridas, E.: Bounds on real eigenvalues and singular values of interval matrices. SIAM J. Matrix Anal. Appl. **31**(4), 2116–2129 (2010)
14. Jaulin, L., Henrion, D.: Contracting optimally an interval matrix without loosing any positive semi-definite matrix is a tractable problem. Reliab. Comput. **11**(1), 1–17 (2005)
15. Kearfott, R.B.: Rigorous Global Search: Continuous Problems. Kluwer, Dordrecht (1996)
16. Kolev, L.V.: Outer interval solution of the eigenvalue problem under general form parametric dependencies. Reliab. Comput. **12**(2), 121–140 (2006)
17. Kolev, L.V.: Determining the positive definiteness margin of interval matrices. Reliab. Comput. **13**(6), 445–466 (2007)
18. Kolev, L.V.: Eigenvalue range determination for interval and parametric matrices. Int. J. Circuit Theory Appl. **38**(10), 1027–1061 (2010)
19. Kreinovich, V., Lakeyev, A., Rohn, J., Kahl, P.: Computational Complexity and Feasibility of Data Processing and Interval Computations. Kluwer (1998)
20. Leng, H.: Real eigenvalue bounds of standard and generalized real interval eigenvalue problems. Appl. Math. Comput. **232**, 164–171 (2014)
21. Liu, W.: Necessary and sufficient conditions for the positive definiteness and stability of symmetric interval matrices. In: Proceedings of the 21st Annual International Conference on Chinese Control and Decision Conference, CCDC 2009, Piscataway, NJ, USA, pp. 4574–4579. IEEE Press (2009)
22. Matcovschi, M.H., Pastravanu, O., Voicu, M.: Right bounds for eigenvalue ranges of interval matrices - estimation principles vs global optimization. Control Eng. Appl. Inform. **14**(1), 3–13 (2012)
23. Mayer, G.: An Oettli-Prager-like theorem for the symmetric solution set and for related solution sets. SIAM J. Matrix Anal. Appl. **33**(3), 979–999 (2012)
24. Meyer, C.D.: Matrix Analysis and Applied Linear Algebra. SIAM, Philadelphia (2000)
25. Mönnigmann, M.: Fast calculation of spectral bounds for hessian matrices on hyperrectangles. SIAM J. Matrix Anal. Appl. **32**(4), 1351–1366 (2011)
26. Nesterov, Y., Nemirovskii, A.: Interior-Point Polynomial Algorithms in Convex Programming. SIAM, Philadelphia (1994)
27. Neumaier, A.: Complete search in continuous global optimization and constraint satisfaction. Acta Numer. **13**, 271–369 (2004)
28. Poljak, S., Rohn, J.: Checking robust nonsingularity is NP-hard. Math. Control Signals Syst. **6**(1), 1–9 (1993)
29. Popova, E.D.: Strong regularity of parametric interval matrices. In: Dimovski, I. et al. (eds.) Mathematics and Education in Mathematics, Proceedings of the 33rd Spring Conference of the Union of Bulgarian Mathematicians, Borovets, Bulgaria, pp. 446–451. BAS (2004)
30. Popova, E.D.: Explicit description of AE solution sets for parametric linear systems. SIAM J. Matrix Anal. Appl. **33**(4), 1172–1189 (2012)

31. Popova, E.D., Hladík, M.: Outer enclosures to the parametric AE solution set. Soft. Comput. **17**(8), 1403–1414 (2013)
32. Rex, G., Rohn, J.: Sufficient conditions for regularity and singularity of interval matrices. SIAM J. Matrix Anal. Appl. **20**(2), 437–445 (1998)
33. Rohn, J.: Positive definiteness and stability of interval matrices. SIAM J. Matrix Anal. Appl. **15**(1), 175–184 (1994)
34. Rohn, J.: VERSOFT: Verification software in MATLAB/INTLAB. Version 10 (2009)
35. Rohn, J.: A handbook of results on interval linear problems. Technical Report No. 1163, Institute of Computer Science, Academy of Sciences of the Czech Republic, Prague (2012)
36. Rohn, J.: A manual of results on interval linear problems. Technical Report No. 1164, Institute of Computer Science, Academy of Sciences of the Czech Republic, Prague (2012)
37. Rump, S.M.: INTLAB - INTerval LABoratory. In: Csendes, T. (ed.) Developments in Reliable Computing, pp. 77–104. Kluwer Academic Publishers, Dordrecht (1999)
38. Shao, J., Hou, X.: Positive definiteness of Hermitian interval matrices. Linear Algebra Appl. **432**(4), 970–979 (2010)
39. Skjäl, A., Westerlund, T.: New methods for calculating αBB-type underestimators. J. Glob. Optim. **58**(3), 411–427 (2014)
40. Vandenberghe, L., Boyd, S.: Semidefinite programming. SIAM Rev. **38**(1), 49–95 (1996)
41. Zimmer, M., Krämer, W., Popova, E.D.: Solvers for the verified solution of parametric linear systems. Comput. **94**(2–4), 109–123 (2012)

Automatic Loop-Shaping of H_∞/μ Problem in QFT Using Interval Consistency Based Hybrid Optimization

R. Jeyasenthil, P.S.V. Nataraj and Harsh Purohit

1 Introduction

Quantitative feedback theory (QFT) [1] is a frequency-domain method of robust control. A key step in QFT is one of synthesizing the controller using loop-shaping method. The loop-shaping is a graphical method to design a controller. In this step, a controller is designed by adding the poles and/or zeros along with gain until the nominal loop transmission function satisfies the performance specification constraint at each frequency. Traditionally, the manual loop-shaping depends on the designer experience and skill, so automatic loop-shaping (ALS) is preferred. It offers the possibility of finding a controller faster and better than the manual one. Existing methods [2–4] attempt to solve this nonlinear and nonconvex problem using convex (or) linear programming techniques, which lead to conservative designs.

An ALS based on reliable deterministic global optimization (namely, interval branch and bound) is proposed in [5]. To speed up this, a method based on hybrid optimization with geometric constraint propagation idea is presented in [6]. These methods, for the first time in the literature, find a global optimum for a particular chosen loop structure [7]. Recently, the QFT controller optimization problem has been formulated as an Interval Constraint Satisfaction Problem (ICSP) with performance specification inequality as constraints [8, 9]. This ICSP formulation uses interval consistency technique (Hull and Box Consistency) to remove the inconsistent values which are not part of the solution [10]. The ICSP formulation gives all the feasible controllers solution in which optimal one is picked up manually based on objective function (e.g. minimum high frequency gain). The main drawback of this ICSP formulation is the computational demand of its search for all feasible solutions.

R. Jeyasenthil (✉) · P.S.V. Nataraj · H. Purohit
IDP in Systems and Control Engineering, Indian Institute of Technology (IIT),
Bombay, India
e-mail: jeya@sc.iitb.ac.in

P.S.V. Nataraj
e-mail: nataraj@sc.iitb.ac.in

H. Purohit
e-mail: harsh.purohit@sc.iitb.ac.in

© Springer International Publishing AG 2018
M. Ceberio and V. Kreinovich (eds.), *Constraint Programming and Decision Making: Theory and Applications*, Studies in Systems, Decision and Control 100,
DOI 10.1007/978-3-319-61753-4_12

2   Basics of QFT

Consider the uncertain linear time-invariant plant given by $P(s) \in \{P(s, \lambda) : \lambda \in \boldsymbol{\lambda}\}$, where $\lambda = (\lambda_1, \lambda_2, ...\lambda_l) \in \mathbf{R}^l$ is a vector of plant parameters. It varies over a box
$$\boldsymbol{\lambda} = \{\lambda \in \mathbf{R}^l : \lambda_i \in [\underline{\lambda_i}, \overline{\lambda_i}], \underline{\lambda_i} < \overline{\lambda_i}, i = 1, ...l\}.$$

The set of plant frequency responses at a given frequency ω, $P(\omega) = \{P(s = j\omega, \lambda) : \lambda \in \boldsymbol{\lambda}\}$ defines a region in the Nichols chart, called the template of the plant at ω.

The nominal open-loop transmission function is defined as $L_0(s, \lambda_0) = C(s) P(s, \lambda_0)$. At each design frequency ω_i, $(i = 1, 2, ..., n)$, robust performance and stability specifications on closed loop system are transformed into bounds in the nominal open-loop transmission function using a corresponding quadratic inequalities [1]. The bound at ω_i is denoted as $B_i(\omega_i)$ and its magnitude varies with the phase of L_0 i.e. $(\angle L_0(j\omega_i))$

Next, a fixed structure controller $C(s)$ is synthesized using loop shaping the nominal loop transfer function. This ensures that the bound constraints, at each ω_i, are respected and nominal closed-loop system is stable. The prefilter $F(s)$ is designed to satisfy the robust tracking specifications such that the closed-loop system should lie in between the lower and upper tracking specifications.

3 H_∞/μ Robust Performance Problem

H_∞ control is one of the robust control method which aims to minimize the ∞ norm of the different sensitivity functions. It solves the control problem as a mathematical optimization problem i.e.

$$J^2(\omega) = (|W_1(j\omega)S_0(j\omega)|^2 + |W_2(j\omega)T_0(j\omega)|^2), \inf_{g \in G} \sup_{\omega} J(\omega) < 1 \qquad (1)$$

where, W_1 and W_2 are frequency-dependent weighting functions in $\Re H_\infty$. Here S_0, T_0 are nominal sensitivity, nominal complementary sensitivity. G is the set of all stabilizing controllers [11]. Simultaneous achievement of robust performance and robust stability is the main objective of any robust control method. Equation (1) is necessary but not sufficient for robust performance (RP). Using structured singular value (μ), a necessary and sufficient condition for robust performance and robust stability is achieved. H_∞/μ robust performance as [11]

$$\mu(\omega) = (|W_1(j\omega)S_0(j\omega)| + |W_2(j\omega)T_0(j\omega)|), \inf_{g \in G} \sup_{\omega} \mu(\omega) < 1 \qquad (2)$$

H_∞/μ robust performance problem is solved in QFT loopshaping framework using the proposed algorithm. The problem (with control effort constraint) considered here is given below

$$\inf_{g \in G} \sup_\omega \mu(\omega) < 1 \tag{3}$$

where

$$\mu(\omega) = (|W_1(j\omega)S_0(j\omega)| + |W_2(j\omega)T_0(j\omega)| + |W_3(j\omega)KS_0(j\omega)|) \leq 1 \tag{4}$$

and $KS_0 =$ Nominal Input Sensitivity function (control effort constraint).

The robust performance problem is converted into QFT bounds using the below quadratic inequality

$$r^2(1 - b^2) + 2r(\cos(\phi + \theta) - ab - bc * \cos(2\phi + \theta)) + (1 - a^2 - c^2 - 2ac) \geq 0 \tag{5}$$

where $r = |L_0|$, $a = |W_1|$, $b = |W_2|$, $c = |W_3|$, ϕ, θ are controller and plant phases respectively.

4 H_∞/μ Problem as QFT Loopshaping: Problem Statement

The QFT controller synthesis problem can be posed as (nonlinear and noncon-vex) constrained optimization problem. The usual objective is to minimize the High Frequency (HF) gain of the controller, which satisfies all the QFT bound constraints. The controller HF gain minimization tends to reduce the amplification of the sensor noise in the HF range (cost of feedback [1]). Usually the performance specifications are at the Low Frequency (LF) range, this motivate us to look for different optimality criteria. So, the objective function considered here is to minimize the magnitude of the nominal open loop transmission function i.e. $|L_0|$ at $\forall \omega_i$. The constraint set C(x) is nonlinear and non-convex and is given by $\mathbf{C_i(x)} = c_i(x)$, where $c_i(x)$ represents a single-valued bound constraint at each design frequency ω_i (either a single-valued upper bound constraint $c_i^u(x)$ or a single-valued lower bound constraint $c_i^l(x)$).

The H_∞/μ problem using QFT synthesis can be formulated as the following constrained global optimization problem.

Minimize

$$\sum_{i=1}^n |L_0(\omega_i)|, \forall \omega_i, i = 1, 2.., n. \tag{6}$$

Subject to

$$c_i^u(x) = |L_0(j\omega_i, x)| - B_i(\angle L_0(j\omega_i, x), \omega_i) \geq 0 \tag{7}$$

$$c_i^l(x) = B_i(\angle L_0(j\omega_i, x), \omega_i) - |L_0(j\omega_i, x)| \geq 0 \tag{8}$$

$$\angle L_0(j\omega_i, x) \geq \psi_{UHFB} \tag{9}$$

Here,

$B_i(\angle L_0(j\omega_i, x), \omega_i)$ = QFT Bound values at each design frequency ω_i (Obtained from MATLAB Toolbox (Borghesani et al. 1995)).

ψ_{UHFB} = Universal High Frequency Bound angle for multiple valued bound (stability margin bound).

5 Proposed Algorithm

In contrast to the above interval methods for automatic controller design, in this paper, an efficient method is proposed. The proposed method combines hybrid optimization and Hull Consistency (HC-4) techniques [10]. The hybrid optimization part combines nonlinear local optimization with interval global optimization methods.

In HC-4 technique, we introduced the bound condition which allows the constraints (to HC-4) that are most suitable for pruning the variable domain. There is no need to input all the constraints to the consistency algorithm because some (or many) of the constraints might have satisfied the relations already. This is ensured by the "bound condition" so processing again these constraints, at every iteration, is not needed for pruning. As a result, the new algorithm is much better, since it does not need to process these constraints on every step.

The pruned box is processed by the local constrained optimization solvers. The local solvers quickly locates the approximate global minimum, provided that updated interval is a feasible one. Here the updated interval is formed by the local solution.

6 Design Example

Consider the uncertain plant transfer function

$$P(s) = \frac{k}{(s+a)(s+b)}; k \in [1, 10], a = [1, 5], b = [0.2, 1];$$

where the nominal parameters are $k = 10, a = 0.2, b = 1$;
Performance Specifications:

- Robust Performance:

$$W1(s) = \frac{(s+5)(s+7)}{1.5(s+0.01)(s+0.1)};$$

- Robust stability:

$$W2 = 0.875;$$

- Control Effort:

$$W3 = 0.001;$$

The controller structure is selected as three poles $(p1, p2, p3)$, two zeros $(z1, z2)$ and gain(k). Initial Controller Parameter values:

$$[k, p1, p2, p3, z1, z2] = [10, 15000], [1, 7.5], [1, 2000], [1, 3850], [10, 50], [100, 500]$$

The design frequency set:

$$\Omega = [0.001, 0.01, 0.05, 0.1, 0.5, 1, 2, 2.5, 3, 5, 6, 10, 50].$$

The tolerance is 10^{-2}. The designed QFT Controller using the proposed algorithm:

$$G(s) = 3096 \frac{(\frac{s}{10} + 1)(\frac{s}{100} + 1)}{(\frac{s}{7.5} + 1)(\frac{s}{2000} + 1)(\frac{s}{3750} + 1)}$$

The H-infinity controller is designed using *hinfsyn* in MATLAB:

$$G_{Hinf}(s) = \frac{2.054 * 10^{10}s^3 + 6.326 * 10^{10}s^2 + 5.044 * 10^{10}s + 7.722 * 10^9}{s^4 + 3.171 * 10^7 s^3 + 5.192 * 10^9 s^2 + 4.114 * 10^{11}s + 7.722 * 10^{11}}$$

The controller gain reduction of 40 dB is possible at high frequency with QFT based controller than H_∞/μ controller.

Performance comparison of the proposed algorithm:

Algorithm	Iterations	Time(sec)
HC4	>1000	–
Moore Skelboe	102	28
Proposed Algorithm	34	17

As compared with existing interval methods [5, 9], it's found that speedup of 40 percent is achieved using the proposed algorithm.

7 Conclusion

The proposed algorithm is applied to automatic QFT loop-shaping of H_∞/μ robust performance problem. The H_∞/μ robust performance problem is converted into QFT loop shaping problem. The advantage of converting is to work with loop transmission function (as in QFT) than sensitivity function (H_∞), which is insensitive to sensor noise problem at high frequency. We extend this problem formulation by adding the input sensitivity function to the original mixed sensitivity problem.

For comparison purpose, the H_∞ control is designed and it's performance is compared with QFT-based H_∞/μ controller. It's observed that, at least for this problem, the QFT based controller gives much better response than the H_∞ controller alone.

References

1. Horowitz, I.M.: Quantitative Feedback Design Theory. QFT Publications, Boulder, CO (1993)
2. Bryant, G.F., Halikias, G.D.: Optimal loop-shaping for systems with large parameter uncertainty via linear programming. Int. J. Control **62**, 557–568 (1995)
3. Chait, Y., Chen, Q., Hollot, C.V.: Automatic loop-shaping of QFT controllers via linear programming. ASME J. Dyn. Syst. Meas. Control **121**, 351–357 (1999)
4. Garcia-Sanz, M., Guillen, J.C.: Automatic loop-shaping of QFT robust controllers via genetic algorithms. In: Proceedings of the 3rd IFAC Symposium on Robust Control Design (2000)
5. Nataraj, P.S.V., Tharewal, S.: An interval analysis algorithm for automated controller synthesis in QFT designs. ASME J. Dyn. Syst. Meas. Control **129**, 311–321 (2007)
6. Nataraj, P.S.V., Kubal, N.: Automatic loop shaping in QFT using hybrid optimization and constraint propagation techniques. Int. J. Robust Nonlinear Control **17**, 251–264 (2007)
7. Cervera, J., Baños, A.: QFT loop shaping with fractional order complex pole-based terms. J. Vib. Control **19**, 294–308 (2012)
8. Deshpande, M.M., Nataraj, P.S.V.: Automated synthesis of fixed structure QFT controller using interval constraint satisfaction techniques. In: Proceeding of the 17th IFAC World Congress, pp. 4976–4981 (2008)
9. Patil, M.D., Nataraj, P.S.V.: Automated synthesis of multivariable QFT controller using interval constraint satisfaction techniques. J. Process Control **22**, 751–765 (2012)
10. Benhamou, F., Goualard, F., Granvilliers, L.: Revising hull and box consistency. In: Proceeding of 16th International Conference on Logic Programming, pp. 230–244 (1999)
11. Nordgren, R.E., Franchek, M.A., Nwokah, O.D.I.: A design procedure for the exact H_∞ SISO-robust performance problem. Int. J. Robust Nonlinear Control **5**, 107–118 (1995)

Existence of the Nash-Optimal Strategies in the Meta-Game

Vyacheslav V. Kalashnikov, Vladimir A. Bulavsky
and Nataliya I. Kalashnykova

1 Introduction

Conjectural variations equilibrium (CVE) was introduced quite long ago as another possible solution concept in static games (*cf.*, [1, 2]). According to this concept, agents behave as follows: each agent chooses his/her most favorable action taking into account that every rival's strategy is a *conjectured function* of his/her own strategy.

In monograph [3], a new concept of conjectural variations equilibrium (CVE) was introduced and investigated, in which the conjectural variations (represented via the so called *influence coefficients* of each agent) provided a new equilibrium paradigm distinct from the Cournot-Nash equilibrium.

The detailed story of the highs and lows of the CVE concept is described in [4]. The main obstacle on the way of admitting this concept is the difficulty of checking its consistency. The *consistency* (or, sometimes, "rationality") of the equilibrium is defined as the coincidence between the conjectural best response of each agent and the conjectured reaction function of the same.

The research activity of the first author was funded by the R&D Department of the Tecnológico de Monterrey (ITESM), Campus Monterrey, and by the SEP-CONACYT project CB-2013-01-221676 (Mexico).

V.V. Kalashnikov (✉)
Tecnológico de Monterrey (ITESM), Campus Monterrey,
Av. Eugenio Garza Sada 2501 Sur, 64849 Monterrey, Nuevo León, Mexico
e-mail: kalash@itesm.mx; slavkamx@gmail.com

V.V. Kalashnikov · V.A. Bulavsky
Central Economics and Mathematics Institute (CEMI), Nakhimovsky Pr. 47,
117418 Moscow, Russia
e-mail: lapissa2010@yandex.ru

N.I. Kalashnykova
Universidad Autónoma de Nuevo León (UANL), Av. Universitaria S/N,
66450 San Nicolás de Los Garza, Nuevo León, Mexico
e-mail: nkalash2009@gmail.com

V.V. Kalashnikov
Sumy State University, Rimsky-Korsakov St. 2, Sumy 40007, Ukraine

© Springer International Publishing AG 2018
M. Ceberio and V. Kreinovich (eds.), *Constraint Programming and Decision Making: Theory and Applications*, Studies in Systems, Decision and Control 100,
DOI 10.1007/978-3-319-61753-4_13

95

To cope with a conceptual difficulty arising in many-player models, Bulavsky proposed in 1997 (*cf.*, [5]) a completely new approach applied later to mixed oligopoly models [4]. Consider a group of n producers ($n \geq 2$) of a homogeneous good with the cost functions $f_i(q_i)$, $i = 1, \ldots, n$, where $q_i \geq 0$ is the output by producer i. Consumers' demand is described by a demand function $G = G(p)$, whose argument p is the market clearing price. Active demand D is nonnegative and does not depend upon the price. The equilibrium between the demand and supply for a given price p is guaranteed by the following balance equality

$$\sum_{i=1}^{n} q_i = G(p) + D. \tag{1}$$

Every producer $i = 1, \ldots, n$, chooses his/her output volume $q_i \geq 0$ so as to maximize his/her profit function

$$\pi_i(p, q_i) = p \cdot q_i - f_i(q_i). \tag{2}$$

Now we postulate that the agents (producers) suppose that their choice of production volumes may affect the price value p. The latter assumption could be defined by a conjectured dependence of the price p upon the output values q_i. If so, the first order maximum condition to describe the equilibrium would have the form:

$$\frac{\partial \pi_i}{\partial q_i} = p + q_i \cdot \frac{\partial p}{\partial q_i} - f_i'(q_i) \begin{cases} = 0, & \text{if } q_i > 0; \\ \leq 0, & \text{if } q_i = 0, \end{cases} \quad \text{for } i = 1, \ldots, n. \tag{3}$$

Thus, we see that to describe the agent's behavior, we need evaluate the behavior of the derivative $\partial p / \partial q_i = -v_i$ rather than the functional dependence of p upon q_i. Then the optimality condition (3) is reduced to

$$\begin{cases} p = v_i q_i + b_i + a_i q_i, & \text{if } q_i > 0; \\ p \leq b_i, & \text{if } q_i = 0. \end{cases} \tag{4}$$

Definition 1 A collection $(p, q_1, \ldots, q_n)$ is called an *exterior equilibrium state* for given influence coefficients $(v_1, \ldots, v_n)$, if the market is balanced, i.e., equality (1) holds, and for each i the maximum conditions (4) are valid.

We assume the following properties of the model's data.

Assumption 1 The demand function $G = G(p) \geq 0$ is defined for $p \in (0, +\infty)$, being non-increasing and continuously differentiable.

The production costs are assumed to be (strictly) convex quadratic functions:

Assumption 2 For each i, the cost function f_i is quadratic and $f_i(0) = 0$, i.e.,

$$f_i(q_i) = (1/2)a_i q_i^2 + b_i q_i, \quad \text{where} \quad a_i > 0, b_i \geq 0, \ i = 1, \ldots, n. \quad (5)$$

From now on, we are going to consider only the case when the set of really producing participants is fixed (i.e., it *doesn't* depend upon the values v_i of the influence coefficients). To guarantee this feature, we make the assumption listed below.

Assumption 3 For the price value $p_0 = \max\limits_{1 \leq j \leq n} b_j$, the following estimate holds:

$$\sum_{i=1}^{n} \frac{p_0 - b_i}{a_i} < G(p_0). \quad (6)$$

Now we establish the following existence result.

Theorem 4 *Under assumptions 1, 2, and 3, for any $D \geq 0$, $v_i \geq 0$, $i = 1, \ldots, n$, there exists uniquely an exterior equilibrium $(p, q_1, \ldots, q_n)$ depending continuously upon the parameters $(D, v_1, \ldots, v_n)$. The equilibrium price $p = p(D, v_1, \ldots, v_n)$ as a function of these parameters is differentiable with respect to D and v_i, $i = 1, \ldots, n$. Moreover, $p(D, v_1, \ldots, v_n) > p_0$, and*

$$\frac{\partial p}{\partial D} = \frac{1}{\displaystyle\sum_{i=1}^{n} \frac{1}{v_i + a_i} - G'(p)}. \quad (7)$$

Now having formula (7) in mind and following the ideas of [2], we introduce the following

Consistency Criterion.
At an exterior equilibrium $(p, q_1, \ldots, q_n)$, the influence coefficients $v_k, k = 1, \ldots, n$, are referred to as *consistent* if the equalities below hold:

$$v_k = \frac{1}{\displaystyle\sum_{i=1, i \neq k}^{n} \frac{1}{v_i + a_i} - G'(p)}, \quad k = 1, \ldots, n. \quad (8)$$

Now we are in a position to define the concept of an interior equilibrium.

Definition 2 A collection $(p, q_1, \ldots, q_n, v_1, \ldots, v_n)$ is called an *interior equilibrium state*, if for the considered influence coefficients, the collection $(p, q_1, \ldots, q_n)$ is an exterior equilibrium state, and the consistency criterion is valid for all $k = 1, \ldots, n$.

The interior equilibrium existence result is as follows:

Theorem 5 *Under assumptions 1, 2, and 3, there exists an interior equilibrium state.*

Theorem 4 allows us to define the following game $\Gamma = (N, V, \Pi, D)$, which will be called the *meta-game*. Here, D is a (fixed) value of the active demand, $N = \{1, \ldots, n\}$ is the set of the same players as in the above-described model, $V = R_+^n$ represents the set of possible strategies, i.e., vectors of conjectures $v = (v_1, \ldots, v_n) \in R_+^n$ accepted by the players, and finally, $\Pi = \Pi(v) = (\pi_1, \ldots, \pi_n)$ is the collection of the payoff values defined (uniquely by Theorem 4) by the strategy vector v.

In general, the Cournot conjectures are not consistent in our single commodity market model. In other words, the Cournot conjectures $v_i = -p'(G)$ usually do not satisfy the (nonlinear) consistency system (8). However, in the meta-game introduced above, the consistent conjectures, determined by (8) provide the Cournot-Nash equilibrium. This result presented below as Theorem 6 was obtained and proved in the previous publication [6].

Theorem 6 *Suppose that Assumptions 1, 2, and 3 hold. Then any Cournot-Nash equilibrium in the meta-game $\Gamma = (N, V, \Pi, D)$ generates a consistent (interior) equilibrium in the original oligopoly. Conversely, every interior (consistent) equilibrium in the original oligopoly is Cournot-Nash equilibrium in the meta-game $\Gamma = (N, V, \Pi, D)$.*

However, since the meta-game strategies set $V = R_+^n$ is unbounded, the existence of at least one Cournot-Nash equilibrium state in this game is by no means easy to check. The following three results (under some extra assumptions) guarantee that the existence of interior equilibrium in the original oligopoly imply the existence of Nash equilibrium in the meta-game. Exactly these three theorems represent the main novelty of this chapter as compared to the previous paper [6]. Since the proofs of Theorems 7, 8, and 9 are quite long, they will be published elsewhere.

Theorem 7 *In addition to Assumptions 1, 2, and 3, suppose that the demand function is affine, that is,*

$$G(p) := \begin{cases} -Kp + T, & \text{if } 0 \leq p \leq \dfrac{T}{K}; \\ 0, & \text{if } p > \dfrac{T}{K}; \end{cases} \tag{9}$$

here, $K > 0, T > 0$. In this case, the consistency criterion for the original oligopoly is the necessary and sufficient condition for the collection of influence conjectures $v = (v_1, \ldots, v_n)$ to be Cournot-Nash equilibrium in the meta-game.

Theorem 8 *Let the assumptions of Theorem 7 be a bit relaxed for the demand function: Instead of (9), suppose that the function G is concave. In addition, if $n = 2$ (duopoly), there exists $\varepsilon > 0$ such that $G'(p) \leq -\varepsilon$ for all $p \geq 0$. Then the consistency criterion for the original oligopoly is the necessary and sufficient condition for the collection of influence conjectures $v = (v_1, \ldots, v_n)$ to be Cournot-Nash equilibrium in the meta-game.*

Since the concavity of the demand function may be a too restrictive requirement, the next theorem relaxes it even more by replacing it with the Lipschitz continuity of the derivative $G'(p)$.

Theorem 9 *Suppose that apart from Assumptions 1, 3, and A??, the regular demand function's derivative is Lipschitz continuous. In more detail, for $n \geq 3$ assume that for any p_1, p_2 the following inequality holds:*

$$|G'(p_1) - G'(p_2)| \leq \frac{1}{2s^2 G(p_0)} |p_1 - p_2|, \tag{10}$$

where $s = \max\{a_1, \ldots, a_n\}$, and the price p_0 is defined in Assumption 3. Next, if $n = 2$ (duopoly), we again suppose that there exists $\varepsilon > 0$ such that $G'(p) \leq -\varepsilon$ for all $p \geq 0$, and the Lipschitz continuity of the demand function is described in the form:

$$|G'(p_1) - G'(p_2)| \leq \frac{2}{\left(\dfrac{a_1 + a_2}{\varepsilon \min\{a_1, a_2\}} + 3s\right)^2 G(p_0)} |p_1 - p_2|, \quad \forall p_1, p_2. \tag{11}$$

Then the consistency criterion for the original oligopoly is the necessary and sufficient condition for the collection of influence conjectures $v = (v_1, \ldots, v_n)$ to be Cournot-Nash equilibrium in the meta-game.

To resume, the paper presents a justification of the concept of consistent conjectures and thus contributes to a better understanding of the nature of conjectural variations equilibrium (CVE). In this paper, we considered an upper level game, in which not the supply volumes q_i but the conjectures (influence coefficients) v_i serve as the players' strategies instead. The remarkable fact we have demonstrated is the following: in the upper level game, the consistent (for the original game) conjectures v_i^* provide for the optimal Cournot-Nash strategies. In other words, if each player i assumes that the other agents stick to their consistent conjectures v_j^*, $j \neq i$, then his/her consistent conjecture v_i^* is optimal for player i, too. The latter means that the vector of conjectures provides the classical Cournot-Nash equilibrium in the upper level game.

References

1. Bowley, A.L.: The Mathematical Groundwork of Economics. Oxford University Press, Oxford (1924)
2. Frisch, R.: Monopole, Polypole–La notion de Force en Économie. Nationaløkonomisk Tidsskrift, vol. 71, pp. 241–259 (1933). (reprinted: Monopoly, Polypoly: The Concept of Force in the Economy. Int. Econ. Papers, vol. 1, pp. 23–36 (1951))
3. Isac, G., Bulavsky, V.A., Kalashnikov, V.V.: Complementarity, Equilibrium Efficiency and Economics. Kluwer Academic Publishers, Dordrecht-London-Boston (2002)

4. Kalashnikov, V.V., Bulavsky, V.A., Kalashnykova, N.I., Castillo, F.J.: Mixed oligopoly with consistent conjectures. European J. Oper. Res. **201**, 729–735 (2011)
5. Bulavsky, V.A.: Structure of demand and equilibrium in a model of oligopoly. Econ. Math. Methods **33**, 112–134 (1997). in Russian
6. Kalashnikov, V.V., Bulavsky, V.A., Kalashnykova, N.I., López-Ramos, F.: Consistent conjectures are optimal Cournot-Nash strategies in the meta-game. Optimization, pp. 1–18 (2016). doi:10.1080/02331934.2016.1238079

Peak-End Rule: A Utility-Based Explanation

Olga Kosheleva, Martine Ceberio and Vladik Kreinovich

1 Peak-End Rule: Description and Need for an Explanation

Peak-end rule: empirical fact. In many situations, people judge their overall experience by the peak and end pleasantness or unpleasantness, i.e., by using only the maximum (minimum) and the last value; see, e.g., [1, 4].

This is true for people's perception of the unpleasantness of a medical procedure, of the quality of the cell phone perception, etc.

Need for an explanation. There is a lot of empirical evidence supporting the peak-end rule, but not much of an understanding. However, at first glance, the rule appears somewhat counter-intuitive: why only peak and last value? why not some average? In this paper, we provide such an explanation based on the traditional decision making theory.

2 Towards an Explanation

Traditional decision making theory: a brief reminder of utility approach. Our objective is to describe the peak-end rule in terms of the traditional decision making theory. According to decision theory, preferences of rational agents can be described in terms of *utility* (see, e.g., [2, 3]): a rational agent selects an action with the largest value of expected utility.

O. Kosheleva (✉) · M. Ceberio · V. Kreinovich
University of Texas at El Paso, El Paso, TX 79968, USA
e-mail: olgak@utep.edu

M. Ceberio
e-mail: mceberio@utep.edu

V. Kreinovich
e-mail: vladik@utep.edu

© Springer International Publishing AG 2018
M. Ceberio and V. Kreinovich (eds.), *Constraint Programming and Decision Making: Theory and Applications*, Studies in Systems, Decision and Control 100,
DOI 10.1007/978-3-319-61753-4_14

Utility is not uniquely defined. Utility is usually defined modulo a linear transformation. In the above experiments, we usually have a fixed *status quo* level which can be taken as 0. Once we fix this value at 0, the only remaining non-uniqueness in describing utility is scaling $u \to k \cdot u$.

Need for a utility-averaging operation. We want to describe the "average" utility corresponding to a sequence of different experiences. We assume that we know the utility corresponding to each moment of time. To get an average utility value, we need to combine these momentous utilities into a single average.

If we have already found the average utility corresponding to two consequent sub-intervals of time, we then need to combine these two averages into a single average corresponding to the whole interval. In other words, we need an operation $a * b$ that, given the average utilities a and b corresponding to two consequent time intervals, generates the average utility of the combined two-stage experience.

Natural properties of the utility-averaging operation.

(1) If we had the same average utility level $a = b$ on both stages, then this same value should be the two-stage average, i.e., we should have $a * a = a$. In mathematical terms, this means that the utility-averaging operation $*$ should be *idempotent*.

(2) If we make one of the stages better, then the resulting average utility should increase (or at least not decrease) as well. In other words, the utility-averaging operation $*$ should be *monotonic* in the sense that if $a \le a'$ and $b \le b'$ then $a * b \le a' * b'$.

(3) Small changes in one of the stages should lead to small changes in the overall average utility; in precise terms, this means that the function $a * b$ must be *continuous*.

(4) For a three-stage situation, with average utilities a, b, and c corresponding to the three stages, we can compute the average utility in two different ways:

- we can first combine the utilities of the first two stages into an average value $a * b$, and then combine this average with c, resulting in $(a * b) * c$;
- alternatively, we can first combine the utilities b and c into $b * c$, and then combine a with $b * c$, resulting in $a * (b * c)$.

The resulting three-stage average should not depend on the order in which we combined the stages, so we should have $(a * b) * c = a * (b * c)$; in mathematical terms, the operation $a * b$ must be *associative*.

(5) Finally, since utility is defined modulo scaling, it is reasonable to require that the utility-averaging operation does not change with scaling:

- In the original scale, we combine a and b and get $a * b$. In the new scale corresponding to a factor $k > 0$, this combined value has the form $k \cdot (a * b)$.
- After re-scaling, the original utilities get the new values $a' = k \cdot a$ and $b' = k \cdot b$. Averaging these two values leads to $a' * b' = (k \cdot a) * (k \cdot b)$ in the new scale.

The resulting average should not depend on how we deduced it, i.e., we should have $(k \cdot a) * (k \cdot b) = k \cdot (a * b)$ for all k, a and b.

What we plan to do. Let us show that the above reasonable requirements largely explain the peak-end phenomenon.

3 Main Result

Proposition 1 *Let $a * b$ be a binary operation on the set of all non-negative numbers which satisfies the following properties:*

(1) it is idempotent, i.e., $a * a = a$ for all a;
(2) it is monotonic, i.e., $a \leq a'$ and $b \leq b'$ imply that $a * b \leq a' * b'$;
(3) it is continuous as a function of a and b;
(4) it is associative, i.e., $(a * b) * c = a * (b * c)$;
(5) it is scale-invariant, i.e., $(k \cdot a) * (k \cdot b) = k \cdot (a * b)$ for all k, a and b.

Then, this operation coincides with one of the following four operations:

- $a_1 * \cdots * a_n = \min(a_1, \ldots, a_n)$;
- $a_1 * \cdots * a_n = \max(a_1, \ldots, a_n)$;
- $a_1 * \cdots * a_n = a_1$;
- $a_1 * \cdots * a_n = a_n$.

Comment. Thus, every utility-averaging operation which satisfies the above reasonable properties means that we select either thse worst or the best or the first or the last utility. This (almost) justifies the peak-end phenomenon, with the only exception that in addition to peak and end, we also have the start $a_1 * \ldots * a_n = a_1$ as one of the options.

Proof

$1°$. For every $a \geq 1$, let us denote $a * 1$ by $\varphi(a)$. For $a = 1$, due to the idempotence, $\varphi(1) = 1 * 1 = 1$. Due to monotonicity, $a \leq a'$ implies that $\varphi(a) \leq \varphi(a')$, i.e., that the function $\varphi(a)$ is (non-strictly) increasing.

$2°$. Due to associativity, for every a, we have $(a * 1) * 1 = a * (1 * 1)$. Due to idempotence, $1 * 1 = 1$, so the above equality takes the form $(a * 1) * 1 = a * 1$, i.e., the form $\varphi(\varphi(a)) = \varphi(a)$. Thus, for every value t from the range of the function $\varphi(a)$ for $a \geq 1$, we have $\varphi(t) = t$.

$3°$. Since the operation $a * b$ is continuous, the function $\varphi(a) = a * 1$ is also continuous. Thus, its range $S \overset{\text{def}}{=} \varphi([1, \infty))$ for $a \in [1, \infty)$ is a connected set, i.e., an interval (finite or infinite). Since the function $\varphi(a)$ is monotonic, and $\varphi(1) = 1$, this interval must start with 1. So, we have three possible options:

- $S = \{1\}$;
- $S = [1, k]$ or $S = [1, k)$ for some $k \in (1, \infty)$;
- $S = [1, \infty)$.

Let us consider these three options one by one.

$3.1°$. In the first case, $\varphi(a) = a * 1 = 1$ for all a. From scale invariance, we can now conclude that for all $a \geq b$, we have $a * b = b \cdot \left(\dfrac{a}{b} * 1\right) = b \cdot 1 = b$.

$3.2°$. In the second case, every value t between 1 and k is a possible value of $\varphi(a)$, thus $\varphi(t) = t * 1 = t$ for all such values t. In particular, for every $\varepsilon > 0$, for the value $t = k - \varepsilon$, we have $\varphi(k - \varepsilon) = k - \varepsilon$. Due to monotonicity, the value $\varphi(k)$ must be not smaller than all these values $k - \varepsilon$, hence not smaller than k. On the other hand, all the values $\varphi(a)$ are less than or equal than k, so we must have $\varphi(k) = k$ as well. Similarly, for values $t \geq k$, due to monotonicity, we have $\varphi(t) \geq k$ and since always $\varphi(t) \leq k$, we conclude that $\varphi(t) = k$ for all $t \geq k$. Now, due to associativity, we have

$$((k - \varepsilon)^2 * (k - \varepsilon)) * 1 = (k - \varepsilon)^2 * ((k - \varepsilon) * 1). \tag{1}$$

Here, due to scale-invariance,

$$(k - \varepsilon)^2 * (k - \varepsilon) = (k - \varepsilon) \cdot ((k - \varepsilon) * 1) = (k - \varepsilon) \cdot \varphi(k - \varepsilon) =$$
$$(k - \varepsilon) \cdot (k - \varepsilon) = (k - \varepsilon)^2, \tag{2}$$

and therefore,

$$((k - \varepsilon)^2 * (k - \varepsilon)) * 1 = (k - \varepsilon)^2 * 1 = \varphi((k - \varepsilon)^2).$$

For $k > 1$, we have $k^2 > k$ and thus, for sufficiently small $\varepsilon > 0$, we have $(k - \varepsilon)^2 > k$. So, $\varphi((k - \varepsilon)^2) = k$, i.e., the left-hand side of the equality (1) is equal to k.

Let us now compute the right-hand side of the equality (1). Here, $(k - \varepsilon) * 1 = k - \varepsilon$ and thus, the right-hand side has the form $(k - \varepsilon)^2 * (k - \varepsilon)$ which, as we already know (Equation (2)), is equal to $(k - \varepsilon)^2$. We already know that the left-hand side is equal to k, and that $(k - \varepsilon)^2 > k$. Thus, the equality (1) cannot be satisfied. This proves that the second case is impossible.

$3.3°$. In the third case, every value $t \geq 1$ is a possible value of $\varphi(a)$, thus

$$\varphi(t) = t * 1 = t$$

for all values $t \geq 1$. Thus, for all $a \geq b$, we have $a * b = b \cdot \left(\dfrac{a}{b} * 1\right) = b \cdot \dfrac{a}{b} = a$.

$4°$. Due to Part 3 of this proof, we have one of the following two cases:

$\geq_1$: for all $a \geq b$, we have $a * b = b$;
$\geq_2$: for all $a \geq b$, we have $a * b = a$.

Similarly, by considering $a \leq b$, we conclude that in this case, we also have two possible cases:

$\leq_1$: for all $a \leq b$, we have $a * b = b$;
$\leq_2$: for all $a \leq b$, we have $a * b = a$.

By combining each of the $\geq$ cases with each of the $\leq$ cases, we get the following four combinations:

$\geq_1, \leq_1$: in this case, $a * b = b$ for all a and b, and therefore, $a_1 * \ldots * a_n = a_n$;
$\geq_1, \leq_2$: in this case, $a * b = \min(a, b)$ for all a and b, and therefore,

$$a_1 * \ldots * a_n = \min(a_1, \ldots, a_n);$$

$\geq_2, \leq_1$: in this case, $a * b = \max(a, b)$ for all a and b, and therefore,

$$a_1 * \ldots * a_n = \max(a_1, \ldots, a_n);$$

$\geq_2, \leq_2$: in this case, $a * b = a$ for all a and b, and therefore, $a_1 * \ldots * a_n = a_1$.

The proposition is proven.

Case of negative utilities. The above formula shows how to combine positive experiences. A similar result can be proven for situations in which we need to combine unpleasant experiences, i.e., experience corresponding to negative utilities; the proof of this result is similar to the proof of Proposition 1.

Remaining open problems. Following the psychological experiments, we only considered the case when all experiences are positive and the case when all experiences are negative. What happens in the general case? If we impose an additional requirement of shift-invariance $(a + u_0) * (b + u_0) = a * b + u_0$, then we can get a result similar to Proposition 1 for this general case as well. But what if we do not impose this additional requirement?

Are all five conditions in Proposition 1 necessary? Some are necessary:

(1) $a * b = a + b$ satisfies all the conditions except for idempotence;
(4) $a * b = \dfrac{a + b}{2}$ satisfies all the conditions except for associativity;
(5) the operation $a * b$ that returns the value from the interval $[\min(a, b), \max(a, b)]$ which is the closest to 1 satisfies all the conditions except for scale invariance.

However, it is not clear whether monotonicity and continuity are needed to prove our results.

Comment. In analyzing the need for these conditions, it may help to know that the set $\{z : z * 1 = z\}$ is a semigroup: indeed, if $z_1 * 1 = z_1$ and $z_2 * 1 = z_2$, then $(z_1 \cdot z_2) * (z_1 * 1) = (z_1 \cdot z_2) * z_1 = z_1 \cdot (z_2 * 1) = z_1 \cdot z_2$ and $((z_1 \cdot z_2) * z_1) * 1 = (z_1 \cdot z_2) * 1$, so associativity implies that $(z_1 \cdot z_2) * 1 = z_1 \cdot z_2$.

Acknowledgements This work was supported in part by the National Science Foundation grants HRD-0734825 and HRD-1242122 (Cyber-ShARE Center of Excellence) and DUE-0926721, by Grants 1 T36 GM078000-01 and 1R43TR000173-01 from the National Institutes of Health, and by a grant N62909-12-1-7039 from the Office of Naval Research.

References

1. Kahneman, D.: Objective happiness. In: Kahneman, D., Diener, E., Schwarz, N. (eds.) Well-Being: The Foundations of Hedonistic Psychology, pp. 3–25. Russell Sage, New York (1999)
2. Luce, R.D., Raiffa, R.: Games and Decisions: Introduction and Critical Survey. Dover, New York (1989)
3. Raiffa, H.: Decision Analysis. Addison-Wesley, Reading, Massachusetts (1970)
4. McFadden, D.L.: The new science of pleasure, national bureau of economic research. Working Paper, No. 18687 (2013)

Similarity Approach to Defining Basic Level of Concepts Explained from the Utility Viewpoint

Joe Lorkowski and Martin Trnecka

1 Formulation of the Problem

What are basic level concepts and why they are important. With the development of new algorithms and faster hardware, computer systems are getting better and better in analyzing images. Computer-based systems are not yet perfect, but in many cases, they can locate human beings in photos, select photos in which a certain person of interest appears, and perform many other practically important tasks.

In general, computer systems are getting better and better in performing well-defined image understanding tasks. However, such systems are much less efficient in more open-ended tasks, e.g., when they need to describe what exactly is described by a photo.

For example, when we present, to a person, a photo of a dog and ask: "What is it?", most people will say "It is a dog". This answer comes natural to us, but, somewhat surprisingly, it is very difficult to teach this answer to a computer. The problem is that from the purely logical viewpoint, the same photo can be characterized on a more abstract level ("an animal", "a mammal") or on a more concrete level ("German Shepherd"). In most situations, out of many possible concepts characterizing a given object, concepts of different levels of generality, humans select a concept of a certain intermediate level. Such preferred concepts are known as *basic level* concepts.

We need to describe basic level concepts in precise terms. Detecting basic level concepts is very difficult for computers. The main reason for this difficulty is that computers are algorithmic machines. So, to teach computers to recognize basic level

J. Lorkowski (✉)
Department of Computer Science, University of Texas at El Paso, ElPaso, TX 79968, USA
e-mail: lorkowski@computer.org

M. Trnecka
Department of Computer Science, Palacky University, Olomouc, Czech Republic
e-mail: martin.trnecka@gmail.com

© Springer International Publishing AG 2018
M. Ceberio and V. Kreinovich (eds.), *Constraint Programming and Decision Making: Theory and Applications*, Studies in Systems, Decision and Control 100,
DOI 10.1007/978-3-319-61753-4_15

concepts, we need to provide an explanation this notion in precise terms—and we are still gaining this understanding.

Current attempts to describe basic level concepts in precise terms: a brief description. When we see a picture, we make a decision which of the concepts to select to describe this picture. In decision making theory, it is known that a consistent decision making can be described by *utility theory*, in which to each alternative A, we put into correspondence a number $u(A)$ called its *utility*—in such a way that a utility of a situation in which we have alternatives A_i with probabilities p_i is equal to $\sum p_i \cdot u(A_i)$; see, e.g., [4, 5, 8, 10, 12].

Naturally, researchers tried to use utility theory to explain the notion of basic level concepts; see, e.g., [3, 6, 7, 14]. In this approach, researchers analyze the effect of different selections on the person's behavior, and come up with the utility values that describes the resulting effects. The utility-based approach describes the basic level concepts reasonably well, but not perfectly. Somewhat surprisingly, a different approach—called *similarity approach*—seems to be more adequate in describing basic level concepts. The idea behind this approach was proposed in informal terms in [13] and has been described more formally in [11]. Its main idea is that in a hierarchy of concepts characterizing a given object, a basic level concept is the one for which the degree of similarity between elements is much higher than for the more abstract (more general) concepts and slightly smaller than for the more concrete (more specific) concepts. For example, we select a dog as a basic level concept because the degree of similarity between different dogs is much larger than similarity between different mammals—but, on the other hand, the degree of similarity between different German Shepherds is not that much higher than the degree of similarity between dogs of various breeds.

In our papers [1, 2], we transformed somewhat informal psychological ideas into a precise algorithms and showed that the resulting algorithms are indeed good in detecting basic level concepts.

Challenging question. From the pragmatic viewpoint, that we have an approach that works well is good news. However, from the methodological viewpoint, the fact that a heuristic approach works better than a well-founded approach based on decision theory—which describes rational human behavior—is a challenge.

What we do in this paper: main result. In this paper, we show—on the qualitative level—that the problem disappears if we describe utility more accurately: under this more detailed description of utility, the decision-making approach leads to the above-mentioned similarity approach.

What we do in this paper: auxiliary result. It is usually more or less clear how to define degree of similarity—or, equivalent, degree of dissimilarity ("distance" $d(x, y)$) between two objects. There are several possible approaches to translate this distance between *objects* into distance between *concepts* (classes of objects). We can use worst-case distance $d(A, B)$ defined as the maximum of all the values $d(x, y)$ for all $x \in A$ and $y \in B$. Alternatively, we can use average distance as the arithmetic average of all the corresponding values $d(x, y)$. In [1], we compared these alternatives; it turns out that the average distance leads to the most adequate description of the basic level concepts.

In this paper, we provide a (qualitative) explanation of this empirical fact as well.

2 Analysis of the Problem and the Resulting Solution

What is the utility associated with concepts of different levels of generality. In the ideal world, when we make a decision in a certain situation, we should take into account all the information about this situation, and we should select the best decision based on this situation.

In practice, our ability to process information is limited. As a result, instead of taking into account all possible information about the object, we use a word (concept) to describe this notion, and then we make a decision based only on this word: e.g., a tiger or a dog. Instead of taking into account all the details of the fur and of the face, we decide to run away (if it is a tiger) or to wave in a friendly manner (if it is a dog).

In other words, instead of making an optimal decision for each object, we use the same decision based on an "average" object from the corresponding class. Since we make a decision without using all the information, based only on an approximate information, we thus lose some utility; see, e.g., [9] for a precise description of this loss.

From this viewpoint, the smaller the classes, the less utility we lose. This is what was used in the previous utility-based approaches to selecting basic level concepts.

However, if the classes are too small, we need to store and process too much information—and the need to waste resources (e.g., time) to process all this additional information also decreases utility. For example, instead of coming up with strategies corresponding to a few basic animals, we can develop separate strategies for short tigers, medium size tigers, larger tigers, etc.—but this would take more processing time and use memory resources which may be more useful for other tasks. While this is a concern, we should remember that we have billions of neurons, enough to store and process huge amounts of information, so this concern is rather secondary in comparison with a different between being eaten alive (if it is a tiger) or not (if it is a dog).

How to transform the above informal description of utility into precise formulas and how this leads to the desired explanations. The main reason for *disutility* (loss of utility) is that in a situation when we actually have an x, we use an approach which is optimal for a similar (but slightly different) object y. For example, instead of making a decision based on observing a very specific dog x, we ignore all the specifics of this dog, and we make a decision based only one the fact that x is a dog, i.e., in effect, we make a decision based on a "typical" dog y.

The larger the distance $d(x, y)$ between the objects x and y, the larger this disutility U. Intuitively, different objects within the corresponding class are similar to each other—otherwise they would not be classified into the same class. Thus, the distance $d(x, y)$ between objects from the same class is small. We can therefore expand the dependence of U on $d(x, y)$ in a Taylor series and keep only the first few terms in this dependence. In general, $U = a_0 + a_1 \cdot d + a_2 \cdot d^2 + \cdots$ When the distance is 0, i.e., when $x = y$, there is no disutility, so $U = 0$. Thus, $a_0 = 0$ and the first non-zero term in the Taylor expansion is $U \approx a_1 \cdot d(x, y)$.

Once we act based on the class label ("concept"), we only know that an object belongs to the class, we do not know the exact object within the class. We may have different objects from this class with different probabilities. By the above property of utility, the resulting disutility of selecting a class is equal to the *average* value of the disutility—and is, thus, proportional to the *average distance* $d(x, y)$ between objects from a given class. *This explains why average distance works better then the worst-case distance.*

When we go from a more abstract concept (i.e., from a larger class) to a more specific concept (i.e., to a smaller class of objects), the average distance decreases—and thus, the main part U_m of disutility decreases: $U'_m < U_m$. However, as we have mentioned, in addition to this main part of disutility U_m, there is also an additional secondary (smaller) part of utility $U_s \ll U_m$, which increases when we go to a more specific concept: $U'_s > U_s$.

On the qualitative level, this means the following: if the less general level has a much smaller degree of similarity (i.e., a drastically smaller average distance between the objects on this level), then selecting a concept on this less general level drastically decreases the disutility $U'_m \ll U_m$, and this decrease $U_m - U'_m \gg 0$ overwhelms the (inevitable) increase $U'_s - U_s$ in the secondary part of disutility, so that $U' = U_m + U'_s < U_m + U_s = U$. On the other hand, if the decrease in degree of similarity is small (i.e., $U'_m \approx U_m$), the increase in the secondary part of disutility $U'_s - U_s$ can over-stage the small decrease $U'_m - U_m$.

A basic level concept is a concept for which disutility U' is smaller than for a more general concept U and than for a more specific concept U''. In view of the above, this means that there should be a drastic difference between the degree of similarity U'_m at this level and the degree of similarity U_m at the more general level—otherwise, on the current level, we would not have smaller disutility. Similarly, there should be a small difference between the degree of similarity at the current level U'_m and the degree of similarity U''_m at the more specific level—otherwise, on the current level, we would not have smaller disutility. *This explains the similarity approach in utility terms.*

Acknowledgements We acknowledge support by the Operational Program Education for Competitiveness Project No. CZ.1.07/2.3.00/20.0060 co-financed by the European Social Fund and Czech Ministry of Education; it was partly performed when Martin Trnecka was visiting University of Texas at El Paso.

References

1. Belohlavek, R., Trnecka, M.: Basic level of concepts in formal concept analysis. In: Proceedings of ICFCA 2012, Springer Verlag, Berlin, Heidelberg, pp. 28-44 (2012)
2. Belohlavek, R., Trnecka, M.: Basic level in formal concept analysis: interesting concepts and psychological ramifications. In: Proceedings of the Twenty-Third International Joint Conference on Artificial Intelligence IJCAI 2013, Beijing, China, pp. 1233-1239 (2013)
3. Corter, J.E., Gluck, M.A.: Explaining basic categories: feature predictability and information. Psychol. Bull. **111**(2), 291–303 (1992)
4. Fishburn, P.C.: Utility Theory for Decision Making. John Wiley & Sons Inc., New York (1969)
5. Fishburn, P.C.: Nonlinear Preference and Utility Theory. The John Hopkins Press, Baltimore, Maryland (1988)
6. Fisher, D.H.: Knowledge acquisition via incremental conceptual clustering. Mach. Learn. **2**(2), 139–172 (1987)
7. Gosselin, F., Schyns, P.G.: Why do we SLIP to the basic level? computational constraints and their implementation. Psychol. Rev. **108**(4), 735–758 (2001)
8. Keeney, R.L., Raiffa, H.: Decisions with Multiple Objectives. John Wiley and Sons, New York (1976)
9. Lorkowski, J., Kreinovich, V.: Likert-scale fuzzy uncertainty from a traditional decision making viewpoint: it incorporates both subjective probabilities and utility information. In: Proceedings of the Joint World Congress of the International Fuzzy Systems Association and Annual Conference of the North American Fuzzy Information Processing Society IFSA/NAFIPS 2013, Edmonton, Canada, pp. 525-530, 24-28 June 2013
10. Luce, R.D., Raiffa, R.: Games and Decisions: Introduction and Critical Survey. Dover, New York (1989)
11. Murphy, G.L.: The Big Book of Concepts. MIT Press, Cambridge, Massachusetts (2002)
12. Raiffa, H.: Decision Analysis. Addison-Wesley, Reading, Massachusetts (1970)
13. Rosch, E.: Principles of categorization. In: Rosch, E., Lloyd, B.B. (eds.) Cognition and Categorization, pp. 27–48. Lawrence Erlbaum Associates, Hillsdale, New Jersey (1978)
14. Zeigenfuse, M.D., Lee, M.D.: A comparison of three measures of the association between a feature and a concept. In: Carlson, L., Holscher, C., Shipley, T.F. (eds.) Proceedings of the 33rd Annual Conference of the Cognitive Science Society, pp. 243-248. Austin, Texas (2011)

Comparisons of Measurement Results as Constraints on Accuracies of Measuring Instruments: When can we Determine the Accuracies from These Constraints?

Christian Servin and Vladik Kreinovich

1 Formulation of the Problem

Need to determine accuracies of measurement instruments. Most information comes from measurements. Measurement results are never absolutely accurate: the measurement result $\tilde{x}$ is, in general, different from the actual (unknown) value x of the corresponding quantity; see, e.g., [7]. To properly process data, it is therefore important to know how accurate are our measurements.

Ideally, we would like to know what are the possible values of measurement errors $\Delta x \stackrel{\text{def}}{=} \tilde{x} - x$, and how frequent are different possible values of Δx. In other words, we would like to know the probability distribution on the set of all possible values of the measurement error Δx.

How accuracies are usually determined: by using a second, much more accurate measuring instrument. One usual way to find the desired probability distribution is to have a second measuring instrument which is much more accurate than the one that we want to estimate. In this case, the measurement error $\Delta x_2 = \tilde{x}_2 - x$ of this second instrument is much smaller than $\Delta x = \tilde{x} - x$ and thus, the difference $\tilde{x} - \tilde{x}_2 = (\tilde{x} - x) - (\tilde{x}_2 - x)$ between the two measurement results can serve as a good approximation to the measurement error. From the sample of such differences, we can therefore find the desired probability distribution for Δx.

What if we do not have a more accurate measuring instrument? But what if the measuring instrument whose accuracy we want to estimate is among the best? In this case, we do not have a much more accurate measuring instrument. What can we do in this case?

C. Servin (✉)
Information Technology Department, El Paso Community College, 919 Hunter,
El Paso, TX 79915, USA
e-mail: cservin@gmail.com

V. Kreinovich
Department of Computer Science, University of Texas at El Paso, El Paso,
TX 79968, USA
e-mail: vladik@utep.edu

© Springer International Publishing AG 2018
M. Ceberio and V. Kreinovich (eds.), *Constraint Programming and Decision Making: Theory and Applications*, Studies in Systems, Decision and Control 100,
DOI 10.1007/978-3-319-61753-4_16

In such situations, we can use the fact that there usually, there are *several* measuring instrument of the type that we want to analyze. Due to measurement errors, for the same quantity, these instruments, in general, produce slightly different measurement results. It is therefore desirable to try to extract the information about measurement accuracies from the differences between these measurement results.

Two possible situations. In some cases, we have a stable manufacturing process that produces several practical identical measuring instruments, for which the probability distributions of measurement error are the same. In such cases, all we need to find is this common probability distribution.

In other cases, we cannot ignore the differences between different instruments. In this case, for each individual measuring instrument, we need to find its own probability distribution.

What is known: case of normal distribution. In many practical situations, the measurement error is caused by the joint effect of numerous independent small factors. In such situations, the Central Limit Theorem (see, e.g., [9]) implies that this distribution is close to Gaussian.

A Gaussian distribution is uniquely determined by its mean (bias) and standard deviation σ. When we only know the differences, we cannot determine the bias: it could be that all the measuring instruments have the same bias, and we will never determine that since we only see the differences. Thus, it makes sense to limit ourselves only to the *random component* of the measurement error, i.e., to the measurement error minus its mean value.

For this "re-normalized" measurement error Δx, the mean is 0. So, all we need to determine is the standard deviation σ. These standard deviations can indeed be determined; see, e.g., [4, 8].

Specifically, hen we have two identical independent measuring instruments, with normally distributed measurement errors Δx_1 and Δx_2, then the difference $\widetilde{x}_2 - \widetilde{x}_1$ is also normally distributed, with variance $V = \sigma^2 + \sigma^2 = 2\sigma^2$. Thus, once we experimentally determine the variance V of this observable difference, we can compute the desired variance σ^2 as $\sigma^2 = \dfrac{V}{2}$.

When we have several different measuring instruments, with unknown standard deviations σ_1, σ_2, σ_3, ..., then for each observable difference $\widetilde{x}_i - \widetilde{x}_j$ the variance is equal to $V_{ij} = \sigma_i^2 + \sigma_j^2$. Thus, once we experimentally determine the three variances V_{12}, V_{23}, and V_{13}, we can find the desired standard deviations by solving the corresponding system of three equations with three unknowns: $V_{12} = \sigma_1^2 + \sigma_2^2$, $V_{23} = \sigma_2^2 + \sigma_3^2$, and $V_{13} = \sigma_1^2 + \sigma_3^2$, whose solution is:

$$\sigma_1^2 = \frac{V_{12} + V_{13} - V_{23}}{2}, \quad \sigma_2^2 = \frac{V_{12} + V_{23} - V_{13}}{2},$$

$$\sigma_3^2 = \frac{V_{13} + V_{23} - V_{12}}{2}.$$

Problem: what if distributions are not Gaussian? Empirical analysis of measuring instruments shows that only slightly more than a half of them have Gaussian measurement errors [3, 6]. What happens in the non-Gaussian case? In such cases, sometimes, we simply cannot uniquely reconstruct the corresponding distributions; see, e.g., [8]. In this paper, we explain when such a reconstruction is possible and when it is not possible.

2 Idea: Let us use Moments

Motivation for using moments. As we have mentioned, a Gaussian distribution with zero mean is uniquely determined by its second moment $M_2 = \sigma^2$. This means that all higher moments $M_k \overset{\text{def}}{=} E[(\Delta x)^k]$ are uniquely determined by the value M_2.

In general, we may have values of M_k which are different from the corresponding Gaussian values. Thus, to describe a general distribution, in addition to the second moment, we also need to describe its higher moments as well.

Moments are sufficient to uniquely describe a distribution: reminder. But even if we know all the moments, will it be sufficient to uniquely determine the corresponding probability distribution? The answer is yes, it is possible, and let us provide a simple reminder of why it is possible—and how can we reconstruct the corresponding distribution.

The usual way to represent a probability distribution of a random variable Δx is by describing its probability density function (pdf) $\rho(\Delta x)$. In many situations, it is convenient to use its *characteristic function*

$$\chi(\omega) \overset{\text{def}}{=} E[\exp(\mathrm{i} \cdot \omega \cdot \Delta x)],$$

where $\mathrm{i} \overset{\text{def}}{=} \sqrt{-1}$, i.e.,

$$\chi(\omega) = \int \rho(\Delta x) \cdot \exp(\mathrm{i} \cdot \omega \cdot \Delta x) \, d\Delta x.$$

From the mathematical viewpoint, the characteristic function is the Fourier transform of the pdf, and it is known that we can uniquely reconstruct a function from its Fourier transform (this reconstruction is known as the *inverse Fourier transform*); see, e.g., [1, 2, 5, 10].

On the other hand, if we use Taylor expansion of the exponential function

$$\exp(z) = 1 + z + \frac{z^2}{2!} + \frac{z^3}{3!} + \cdots + \frac{z^k}{k!} + \cdots,$$

then the characteristic functions takes the form

$$\chi(\omega) = E\left[1 + i \cdot \omega \cdot \Delta x - \frac{1}{2!} \cdot \omega^2 \cdot (\Delta x)^2 + \cdots + \frac{i^k}{k!} \cdot \omega^k \cdot (\Delta x)^k + \cdots\right],$$

i.e.,

$$\chi(\omega) = 1 - \frac{1}{2} \cdot \omega^2 \cdot M_2 + \cdots + \frac{i^k}{k!} \cdot \omega^k \cdot M_k + \cdots$$

Thus, if we know all the moments M_k, we can uniquely reconstruct the characteristic function and thus, uniquely reconstruct the desired pdf.

Important fact: for a symmetric distribution, odd moments are zeros. In the following analysis, it is important to use the fact that for a symmetric distribution, i.e., a distribution for which $\rho(-\Delta x) = \rho(\Delta x)$, add odd moments M_{2s+1} are equal to 0:

$$M_{2s+1} = \int \rho(\Delta x) \cdot (\Delta x)^{2s+1} \, d\Delta x.$$

Indeed, if we replace Δx to $\Delta x' \stackrel{\text{def}}{=} -\Delta x$, then $d\Delta x = -d\Delta x'$, $(\Delta x)^{2s+1} = -(\Delta x')^{2s+1}$ and thus, the above integral takes the form

$$M_{2s+1} = -\int \rho(-\Delta x') \cdot (\Delta x')^{2s+1} \, d\Delta x' = -\int \rho(\Delta x') \cdot (\Delta x')^{2s+1} \, d\Delta x',$$

so $M_{2s+1} = -M_{2s+1}$ and hence, $M_{2s+1} = 0$.

3   Case when have Several Identical Measuring Instruments

Description of the case: reminder. In this cases, we have several measuring instruments, with the same probability distribution and thus, with the same moments M_2, M_3, etc. The only available information consists of the differences $\Delta x_1 - \Delta x_2 = \tilde{x}_1 - \tilde{x}_2$. Based on the observations, we can determine the probability distribution for each such difference, and thus, we can determine the moments M'_k of this difference.

   We would like to use these observable moments $M'_k = E[(\Delta x_1 - \Delta x_2)^k]$ to find the desired differences $M_k = E[(\Delta x)^k]$.

What is known: case of second moments. For $k = 2$, we have $M'_2 = 2M_2$ and thus, we can uniquely reconstruct the desired second moment M_2 from the observed second moment M'_2.

Natural next case: third moments. Can we similarly reconstruct the desired third moment $M_3 = E[(\Delta x)^3]$ based on the observed third moment $M'_3 = E[(\Delta x_1 - \Delta x_2)^3]$?
   Here,

$$(\Delta x_1 - \Delta x_2)^3 = (\Delta x_1)^3 - 3 \cdot (\Delta x_1)^2 \cdot \Delta x_2 + 3 \cdot \Delta x_1 \cdot (\Delta x_2)^2 - (\Delta x_2)^3,$$

so, due to linearity of the mean and to the fact that the measurement errors Δx_1 and Δx_2 corresponding to two measuring instruments are assumed to be independent, we conclude that

$$M_3' = E[(\Delta x_1 - \Delta x_2)^3] = E[(\Delta x_1)^3] - 3 \cdot E[(\Delta x_1)^2] \cdot E[\Delta x_2] +$$

$$3 \cdot E[\Delta x_1] \cdot E[(\Delta x_2)^2] - E[(\Delta x_2)^3].$$

In this case, $E[\Delta x_i] = 0$ and $E(\Delta x_1)^3] = E[(\Delta x_2)^3] = M_3$, so

$$M_3' = M_3 - M_3 = 0.$$

In other words, the observed third moment M_3' is always equal to 0, and thus, carries no information about M_3.

So, the only case when we can reconstruct M_3 is when we know it already. One such case is when we know that the distribution is symmetric. In turns out that in this case, we can reconstruct all the moments and thus, we can uniquely reconstruct the original probability distribution.

When the probability distribution of the measurement error is symmetric, this distribution can be uniquely determined from the observed differences. For a symmetric distribution, all odd moments are equal to 0. Thus, to uniquely determine a symmetric distribution, it is sufficient to determine all its even moments M_{2s}. Let us prove, by induction, that we can reconstruct all these even moments.

We already know that we can reconstruct M_2. Let us assume that we already know how to reconstruct the moments $M_2, \ldots, M_{2s}$. Let us show how to reconstruct the next moment $M_{2s+2} = E[(\Delta x)^{2s+2}]$. For this, we will use the observed moment $M_{2s+2}' = E[(\Delta x_1 - \Delta x_2)^{2s+2}]$. Here,

$$(\Delta x_1 - \Delta x_2)^{2s+2} = (\Delta x_1)^{2s+2} - (2s+2) \cdot (\Delta x_1)^{2s+1} \cdot \Delta x_2 +$$

$$\frac{(2s+2) \cdot (2s+1)}{1 \cdot 2} \cdot (\Delta x_1)^{2s} \cdot (\Delta x_2)^2 - \ldots +$$

$$\frac{(2s+2) \cdot (2s+1)}{1 \cdot 2} \cdot (\Delta x_1)^2 \cdot (\Delta x_2)^{2s} - (2s+2) \cdot \Delta x_1 \cdot (\Delta x_2)^{2s+1} + (\Delta x_2)^{2s+2}.$$

Thus,

$$M_{2s+2}' = E[(\Delta x_1)^{2s+2}] - (2s+2) \cdot E[(\Delta x_1)^{2s+1}] \cdot E[\Delta x_2] +$$

$$\frac{(2s+2) \cdot (2s+1)}{1 \cdot 2} \cdot E[(\Delta x_1)^{2s}] \cdot E[(\Delta x_2)^2] - \ldots +$$

$$\frac{(2s+2) \cdot (2s+1)}{1 \cdot 2} \cdot E[(\Delta x_1)^2] \cdot E[(\Delta x_2)^{2s}] -$$

$$(2s + 2) \cdot E[\Delta x_1] \cdot E[(\Delta x_2)^{2s+1}] + E[(\Delta x_2)^{2s+2}],$$

i.e.,

$$M'_{2s+2} = M_{2s+2} + \frac{(2s + 2) \cdot (2s + 1)}{1 \cdot 2} \cdot M_{2s} \cdot M_2 + \ldots +$$

$$\frac{(2s + 2) \cdot (2s + 1)}{1 \cdot 2} \cdot M_2 \cdot M_{2s} + M_{2s+2}.$$

Thus,

$$2M_{2s+2} = M'_{2s+2} - \frac{(2s + 2) \cdot (2s + 1)}{1 \cdot 2} \cdot M_{2s} \cdot M_2 - \ldots -$$

$$\frac{(2s + 2) \cdot (2s + 1)}{1 \cdot 2} \cdot M_2 \cdot M_{2s}.$$

We know the value M'_{2s+2}, and we assumed that we have already shown that we can uniquely determine the moments M_2, ..., M_{2s}. Thus, we can indeed uniquely determine the moment M_{2s+2}.

Induction proves that we can indeed determine all the even moments.

4   Case when have Several Different Measuring Instruments

Description of the case: reminder. In this case, we have several measuring instruments with, in general, different probability distributions. For each of the measuring instruments i, we want to find the corresponding moments

$$M_{k,i} = E[(\Delta x_i)^k].$$

To find these moments, we can use the observe moments

$$M'_{k,i,j} = E[(\Delta x_i - \Delta x_j)^k].$$

What is known: case of second moments. For $k = 2$, we have $M'_{2,i,j} = M_{2,i} + M_{2,j}$, so we can uniquely reconstruct the desired second moments $M_{2,i}$ from the observed moments $M'_{2,i,j}$ by using the following formulas:

$$M_{2,1} = \frac{M'_{2,1,2} + M'_{2,1,3} - M'_{2,2,3}}{2}, \quad M_{2,2} = \frac{M'_{2,1,2} + M'_{2,2,3} - M'_{2,1,3}}{2},$$

$$M_{2,3} = \frac{M'_{2,1,3} + M'_{2,2,3} - M'_{2,1,2}}{2}.$$

Natural next case: third moments. Can we similarly reconstruct the desired third moments $M_{3,i} = E[(\Delta x_i)^3]$ based on the observed third moments $M'_{3,i,j} = E[(\Delta x_i - \Delta x_j)^3]$?

Here,

$$(\Delta x_i - \Delta x_j)^3 = (\Delta x_i)^3 - 3 \cdot (\Delta x_i)^2 \cdot \Delta x_j + 3 \cdot \Delta x_i \cdot (\Delta x_j)^2 - (\Delta x_j)^3,$$

so, due to linearity of the mean and to the fact that the measurement errors Δx_i and Δx_j corresponding to two measuring instruments are assumed to be independent, we conclude that

$$M'_{3,i,j} = E[(\Delta x_i - \Delta x_j)^3] = E[(\Delta x_i)^3] - 3 \cdot E[(\Delta x_i)^2] \cdot E[\Delta x_j] +$$

$$3 \cdot E[\Delta x_i] \cdot E[(\Delta x_j)^2] - E[(\Delta x_j)^3].$$

In this case, $E[\Delta x_i] = E[\Delta x_j] = 0$ and $E(\Delta x_i)^3] = M_{3,i}$, so

$$M'_{3,i,j} = M_{3,i} - M_{3,j}.$$

Since we only know the differences between the their moments, we cannot uniquely reconstruct these moments $M_{3,i}$: for example, if we add a constant to all the values $M_{3,i}$, all the observed differences will not change.

So, the only case when we can reconstruct the third moments $M_{3,i}$ is when we have some information about them already. One such case is when we know that for one of the measuring instruments, the probability distribution of measurement errors is symmetric. In turns out that in this case, we can reconstruct all the moments and thus, we can uniquely reconstruct all the original probability distributions.

When the probability distribution of one of the measurement errors is symmetric, all distributions can be uniquely determined from the observed differences. Without losing generality, let us assume that the probability distribution of the measurement error is symmetric for the 1st measuring instrument. For a symmetric distribution, all odd moments are equal to 0; thus, we have $M_{2s+1,1} = 0$ for all s. Let us prove, by induction, that we can reconstruct all the moments of all the distributions.

We already know that we can reconstruct the second moments $M_{2,i}$. Let us assume that we already know how to reconstruct the moments $M_{2,i}, \ldots, M_{n,i}$. Let us show how to reconstruct the next moments $M_{n+1,i} = E[(\Delta x_i)^{n+1}]$. For this, we will use the observed moments $M'_{n+1,i,j} = E[(\Delta x_i - \Delta x_j)^{n+1}]$. We will consider two cases:

- when n is odd, i.e., $n = 2s + 1$ and $n + 2 = 2s + 2$, and
- when n is even, i.e., $n = 2s$ and $n + 1 = 2s + 1$.

First case. Let us first consider the first case. Here,

$$(\Delta x_i - \Delta x_j)^{2s+2} = (\Delta x_i)^{2s+2} - (2s + 2) \cdot (\Delta x_i)^{2s+1} \cdot \Delta x_j +$$

$$\frac{(2s+2)\cdot(2s+1)}{1\cdot2}\cdot(\Delta x_i)^{2s}\cdot(\Delta x_j)^2-\ldots+$$

$$\frac{(2s+2)\cdot(2s+1)}{1\cdot2}\cdot(\Delta x_i)^2\cdot(\Delta x_j)^{2s}-(2s+2)\cdot\Delta x_i\cdot(\Delta x_j)^{2s+1}+(\Delta x_j)^{2s+2}.$$

Thus,

$$M'_{2s+2,i,j}=E[(\Delta x_i)^{2s+2}]-(2s+2)\cdot E[(\Delta x_i)^{2s+1}]\cdot E[\Delta x_j]+$$

$$\frac{(2s+2)\cdot(2s+1)}{1\cdot2}\cdot E[(\Delta x_i)^{2s}]\cdot E[(\Delta x_j)^2]-\ldots+$$

$$\frac{(2s+2)\cdot(2s+1)}{1\cdot2}\cdot E[(\Delta x_i)^2]\cdot E[(\Delta x_j)^{2s}]-$$

$$(2s+2)\cdot E[\Delta x_i]\cdot E[(\Delta x_j)^{2s+1}]+E[(\Delta x_j)^{2s+2}],$$

i.e.,

$$M'_{2s+2,i,j}=M_{2s+2,i}+\frac{(2s+2)\cdot(2s+1)}{1\cdot2}\cdot M_{2s,i}\cdot M_{2,j}+\ldots+$$

$$\frac{(2s+2)\cdot(2s+1)}{1\cdot2}\cdot M_{2,i}\cdot M_{2s,j}+M_{2s+2,j}.$$

Thus,

$$M_{2s+2,i}+M_{2s+2,j}=s_{ij}\overset{\text{def}}{=}M'_{2s+2,i,j}-\frac{(2s+2)\cdot(2s+1)}{1\cdot2}\cdot M_{2s,i}\cdot M_{2,j}-\ldots-$$

$$\frac{(2s+2)\cdot(2s+1)}{1\cdot2}\cdot M_{2,i}\cdot M_{2s,j}.$$

We know the value $M'_{2s+2,i,j}$, and we assumed that we have already shown that we can uniquely determine the moments $M_{2,i},\ldots,M_{2s+1,i}$. Thus, we can indeed uniquely determine the values $s_{ij}=M_{2s+2,i}+M_{2s+2,j}$.

Based on these values, we can uniquely reconstruct the moments $M_{n+1,i}=M_{2s+2,i}$ as follows:

$$M_{2s+2,1}=\frac{s_{12}+s_{13}-s_{23}}{2},\quad M_{2s+2,2}=\frac{s_{12}+s_{23}-s_{13}}{2},$$

$$M_{2s+2,3}=\frac{s_{13}+s_{23}-s_{12}}{2}.$$

Second case. Let us now consider the second case, when $n = 2s$ and $n + 1 = 2s + 1$. Since we assumed that for the first measuring instrument, the probability distribution is symmetric, we get $M_{2s+1,1} = E[(\Delta x_1)^{2s+1}] = 0$.

For every $i \neq 1$, we have

$$(\Delta x_i - \Delta x_1)^{2s+1} = (\Delta x_i)^{2s+1} - (2s + 2) \cdot (\Delta x_i)^{2s} \cdot \Delta x_1 +$$

$$\frac{(2s + 1) \cdot 2s}{1 \cdot 2} \cdot (\Delta x_i)^{2s-1} \cdot (\Delta x_1)^2 - \ldots +$$

$$\frac{(2s + 1) \cdot 2s}{1 \cdot 2} \cdot (\Delta x_i)^2 \cdot (\Delta x_1)^{2s-1} - (2s + 1) \cdot \Delta x_i \cdot (\Delta x_1)^{2s} + (\Delta x_1)^{2s+1}.$$

Thus,

$$M'_{2s+1,i,1} = E[(\Delta x_i)^{2s+1}] - (2s + 2) \cdot E[(\Delta x_i)^{2s+1}] \cdot E[\Delta x_1] +$$

$$\frac{(2s + 1) \cdot 2s}{1 \cdot 2} \cdot E[(\Delta x_i)^{2s-1}] \cdot E[(\Delta x_1)^2] - \ldots +$$

$$\frac{(2s + 1) \cdot 2s}{1 \cdot 2} \cdot E[(\Delta x_i)^2] \cdot E[(\Delta x_1)^{2s-1}] -$$

$$(2s + 1) \cdot E[\Delta x_i] \cdot E[(\Delta x_1)^{2s}] + E[(\Delta x_1)^{2s+1}],$$

i.e.,

$$M'_{2s+1,i,1} = M_{2s+1,i} + \frac{(2s + 1) \cdot 2s}{1 \cdot 2} \cdot M_{2s-1,i} \cdot M_{2,1} + \ldots +$$

Thus,

$$M_{2s+1,i} = M'_{2s+1,i,1} - \frac{(2s + 1) \cdot 2s}{1 \cdot 2} \cdot M_{2s,i} \cdot M_{2,1} - \ldots .$$

We know the value $M'_{2s+1,i,1}$, and we assumed that we have already shown that we can uniquely determine the moments $M_{2,i}, \ldots, M_{2s,i}$. Thus, we can indeed uniquely determine the moments $M_{n+1,i} = M_{2s+1,i}$.

Conclusion. In both cases, the induction step is proven, so induction proves that we can indeed determine all the moments of all the distributions.

Acknowledgements This work was supported in part by the National Science Foundation grants HRD-0734825 and HRD-1242122 (Cyber-ShARE Center of Excellence).

References

1. Bracewell, R.N.: Fourier Transform and Its Applications. McGraw Hill, New York (1978)
2. Cormen, T.H., Leiserson, C.E., Rivest, R.L., Stein, C.: Introduction to Algorithms. MIT Press, Cambridge, Massachusetts (2009)
3. Novitskii, P.V., Zograph, I.A.: Estimating the Measurement Errors. Energoatomizdat, Leningrad (1991). (in Russian)
4. Ochoa, O., Velasco, A., Servin, C.: Towards model fusion in geophysics: how to estimate accuracy of different models. J. Uncertain Syst. **7**(3), 190–197 (2013)
5. Orfanidis, S.: Introduction to Signal Processing. Prentice Hall, Upper Saddle River, New Jersey (1995)
6. Orlov, A.I.: How often are the observations normal? Ind. Lab. **57**(7), 770–772 (1991)
7. Rabinovich, S.: Measurement Errors and Uncertainties: Theory and Practice. American Institute of Physics, New York (2005)
8. Servin, C., Kreinovich, V.: Propagation of Interval and Probabilistic Uncertainty in Cyberinfrastructure-Related Data Processing and Data Fusion. Springer Verlag, Berlin, Heidelberg (2015)
9. Sheskin, D.J.: Handbook of Parametric and Nonparametric Statistical Procedures. Chapman and Hall/CRC Press, Boca Raton, Florida (2011)
10. Sneddon, I.N.: Fourier Transforms. Dover Publisher, New York (2010)

Dow Theory's Peak-and-Trough Analysis Justified

Chrysostomos D. Stylios and Vladik Kreinovich

1 Formulation of the Problem

Peak-and-trough analysis. In the early 20th century, a theory—known as Dow Theory—was developed for forecasting the behavior of different prices, such as stock prices, equity prices, etc. The main idea behind this theory is that:

- similarly to calculus, where the important first step in the analysis of a function is finding its local minima and maxima,
- the important information about the changes in stock market prices can be obtained if we mark local maxima ("peaks") and local minima ("troughs"); see, e.g., [4, 6].

This analysis is still in use. The resulting peak-and-trough analysis was widely used in the 1920s and early 1930s, until a paper [3] showed the deficiency of the corresponding forecasting techniques.

This paper used then-prevalent expected-return values to analyze the quality of the Dow Theory recommendations. By the 1990s, however, it became clear that when comparing different stock recommendations, it is important to also take into account the corresponding *risks*.

It turns out that if we take risk into account, then the Dow Theory recommendations are not inferior at all, these predictions are actually reasonably good; see, e.g., [7]. As a result, the peak-and-trough analysis has been revived—and it is still used in financial analysis.

C.D. Stylios (✉)
Laboratory of Knowledge and Intelligent Computing, Department of Computer Engineering,
Technological Educational Institute of Epirus, 47100 Kostakioi, Arta, Greece
e-mail: stylios@teiep.gr

V. Kreinovich
Department of Computer Science, University of Texas at El Paso, 500 W. University,
El Paso, TX 79968, USA
e-mail: vladik@utep.edu

Comment. The actual dependence of the stock prices (and other prices) on time t comes with *noise*: random fluctuations caused by many random factors. From the purely mathematical viewpoint, this means that the dependence oscillates all the time, so almost every moment of time has it local minima and local maxima. What the peak-and-trough analysis suggest, of course, is *not* to use all these moments of time, but only to use moments of *true* local minima and maxima, i.e., moments when we can be sure that the local extremum is not cause by the noise itself.

So, to apply this analysis, we need first to be able to distinguish between local extrema which may be due to noise and the real local extrema. There exist efficient algorithms for making this distinction. For example, in situations when all we know about the noise $n(t)$ is that its absolute value $|n(t)|$ is bounded by some value n_0 ($|n(t)| \leq n_0$), there is an efficient (linear time) algorithm for detecting real local extrema; see, e.g., [12].

Similar ideas works well in engineering as well. When we only take into account the local extrema, this means that:

- for all the moments of time between a local maximum and the following local minimum, the value $x(t)$ decreases; we do not have any information about how exactly it decreases, we only know that is decreases;
- similarly, for all the moments of time between a local minimum and the following local maximum, the value $x(t)$ increases; we do not have any information about how exactly it increases, we only know that it increases.

In other words, for each moment of time t, we only have one of the following three pieces of information about how the signal $x(t)$ changes in the small vicinity of this moment t:

- we may know that there is a local extremum in this vicinity; in this case, in this vicinity, the value $x(t)$ practically does not change,
- we may know that the value $x(t)$ decreases in this vicinity,
- or, alternatively, we may know that the value $x(t)$ increases in this vicinity.

Interestingly, many efficient methods of signal compression—starting with the so-called *delta-modulation*—are based on recording, for each moment of time, exactly one of these three situations: 0 (no change), $-$ (decrease), or $+$ (increase); see, e.g., [1, 2, 5, 8, 10, 13].

Why is peak-and-trough analysis efficient? What is not clear is why the peak-and-trough analysis, an analysis that ignores all monotonicity segments and only takes into account the local extrema, is efficient.

Comment. Similarly to financial applications, from the theoretical viewpoint, the engineering-oriented empirical success of delta-modulation techniques is also largely a mystery.

What we do in this paper. In this paper, we provide a possible explanation for the efficiency of peak-and-trough techniques.

2 Our Explanation

In the first approximation, it is reasonable to only process the most important values. Ideally, we should take into account the values $x(t)$ of the stock price at all moments of time t. The problem is that there is a large amount of these data points, and without a clear understanding of the underlying processes, it is difficult to meaningfully process all this data.

It is therefore reasonable, in the first approximation, to only concentrate on the *most important* stock price values and ignore the less important values.

Which values should we take into account? As the stock price fluctuates, it attains different values x. Some values appear more frequently, some values appear more rarely. It therefore makes sense to concentrate on the prices that appear the largest number of times.

Of course, from the practical viewpoint, very close values x can be viewed as identical. So, when we talk about the time that a value x appears, we mean the time when the value $x(t)$ is within an interval $[x - \delta, x + \delta]$ for some small $\delta > 0$.

How to decide which values are most frequent? The values $x(t)$ are rarely stable, then usually change with time. Thus, the time period during which the value is within a given interval $[x - \delta, x + \delta]$ is small. If we had $x(t_0) = x$ for some moment t_0, this means that the neighboring moments of time t at which $x(t) \in [x - \delta, x + \delta]$ are close to t_0, i.e., have the form $t = t_0 + \Delta t$, where $\Delta t \ll t_0$. For such small values Δt, we can ignore quadratic and higher order terms in the dependence of $x(t)$ on t, and use the linear approximation

$$x(t_0 + \Delta t) \approx x(t_0) + x'(t_0) \cdot \Delta t = x + x'(t_0) \cdot \Delta t. \tag{1}$$

Thus, the length ℓ of the time interval during which

$$x(t) = x(t_0 + \Delta t) \in [x - \delta, x + \delta]$$

is equal to

$$\ell = \frac{2\delta}{|x'(t_0)|}. \tag{2}$$

Resulting explanation. We have decided to only consider the values $x(t_0)$ for which this time interval ℓ is large. According to the formula (2), this means that we should only consider the values $x(t_0)$ at the moments t_0 at which the derivative $x'(t_0)$ is close to 0—i.e., only the values in the vicinity of points where the derivative is equal to 0. These points are exactly local minima and local maxima—as well as possible non-minimum and non-maximum stationary points.

Thus, we indeed have an explanation of why the peak-and-trough strategy is successful.

3   Additional Theoretical Confirmation of Our Justification

Another situation where extreme points frequently occur. B. S. Tsirelson noticed
[11] that in many cases, when we reconstruct the signal from the noisy data, and we
assume that the resulting signal belongs to a certain class, the reconstructed signal
is often an *extreme* point from this class. For example:

- when we assume that the reconstructed signal is monotonic, the reconstructed
  function is often (piece-wise) constant;
- if we additional assume that the signal is smooth (one time differentiable, from the
  class C^1), the result is usually one time differentiable but rarely twice differentiable,
  etc.

This situation has an explanation. To explain this phenomenon, Tsirelson provided
the following *geometric* explanation to this fact: namely, when we reconstruct a
signal from a mixture of a signal and a Gaussian noise, then the *maximum likelihood*
estimation (a traditional statistical technique; see, e.g., [9]) means that we look for a
signal that belongs to the priori class, and that is the closest (in the L^2-metric) to the
observed "signal+noise".

   In particular, if the signal is determined by finitely many (say, d) parameters, we
must look for a signal $s = (s_1, \ldots, s_d)$ from the a priori set $A \subseteq R^d$ that is the closest
(in the usual Euclidean sense) to the observed values

$$o = (o_1, \ldots, o_d) = (s_1 + n_1, \ldots, s_d + n_d),$$

where n_i denotes the (unknown) values of the noise.

   Since the noise is Gaussian, we can usually apply the *Central Limit Theorem* [9]
and conclude that the average value of $(n_i)^2$ is close to σ^2, where σ is the standard
deviation of the noise. In other words, we can conclude that

$$(n_1)^2 + \cdots + (n_d)^2 \approx d \cdot \sigma^2.$$

In geometric terms, this means that the distance

$$\sqrt{\sum_{i=1}^{d}(o_i - s_i)^2} = \sqrt{\sum_{i=1}^{d} n_i^2}$$

between s and o is $\approx \sigma \cdot \sqrt{d}$. Let us denote this distance $\sigma \cdot \sqrt{d}$ by ε.

   Let us first, for simplicity, consider the case when $d = 2$, and when A is a convex
polygon. Then, we can divide all points p from the exterior of A that are ε-close to
A into several zones depending on what part of A is the closest to p:

- one of the *sides*, or
- one of the *edges*.

Geometrically, the set of all points for which the closest point $a \in A$ belongs to the *side e* is bounded by the straight lines orthogonal (perpendicular) to e. The total length of this set is therefore equal to the length of this particular side; hence, the total length of all the points that are the closest to all the sides is equal to the *perimeter* of the polygon. This total length thus does not depend on ε at all.

On the other hand, the set of all the points at the distance ε from A grows with the increase in ε; its length grows approximately as the length of a circle, i.e., as const·ε.

When ε increases, the (constant) perimeter is a vanishing part of the total length. Hence, for large ε:

- the fraction of the points that are the closest to one of the sides tends to 0, while
- the fraction of the points p for which the *closest* is one of the *edges* tends to 1.

Similar arguments can be repeated for any dimension. For the same noise level σ, when d increases, the distance $\varepsilon = \sigma \cdot \sqrt{d}$ also increases, and therefore, for large d, for "almost all" observed points o, the reconstructed signal is one of the extreme points of the a priori set A.

A similar explanation can be applied to our case as well. In our case, as we showed in the previous section, extreme values are also much more frequently observed than others. Thus, our argument can be viewed as a particular case of the general geometric explanation proposed by Tsirelson.

Acknowledgements This work was supported in part by the National Science Foundation grants HRD-0734825 and HRD-1242122 (Cyber-ShARE Center of Excellence) and DUE-0926721. This work was performed when C. Stylios was a Visiting Researcher at the University of Texas at El Paso.

The authors are thankful to Djuro Zrilic for valuable discussions.

References

1. Bourdopoulos, G.I., Pnevmatikakis, A.: Delta-Sigma Modulators: Modeling, Design and Applications. Imperial College Press, London, UK (2003)
2. Breems, L., Huijsing, J.: Continuous-Time Sigma-Delta Modulation for A/D Conversion in Radio Receivers. Springer Verlag, New York (2013)
3. Cowles, A.: Can stock market forecasters forecast? Econom. **1**(3), 309–324 (1933)
4. Hamilton, W.P.: The Stock Market Barometer: A Study of its Forecast Value Based on Charles H. Dow's Theory of the Price Movement. Barrons, New York (1922)
5. Janssen, E., van Roermund, A.: Look-Ahead Based Sigma-Delta Modulation. Springer, Dordrecht, Heidelberg, London, New York (2011)
6. Rhea, R.: The Dow Theory. Barron's, New York (1932)
7. Schannep, J.: Dow Theory for the 21st Century: Technical Indicators for Improving Your Investment Results. Wiley, New York (2008)
8. Schreier, R., Temes, G.C.: Understanding Delta-Sigma Data Converters. IEEE Press, Piscataway, New Jersey (2004). Wiley, Hoboken, New Jersey
9. Sheskin, D.J.: Handbook of Parametric and Nonparametric Statistical Procedures. Chapman & Hall/CRC, Boca Raton, Florida (2011)
10. Sira-Ramírez, H.: Sliding Mode Control: The Delta-Sigma Modulation Approach. Birkhäuser, Basel, Switzerland (2015)

11. Tsirel'son, B.S.: A geometrical approach to maximum likelihood estimation for infinite-dimensional Gaussian location. I. Theory Probab. Appl. **27**, 411–418 (1982)
12. Villaverde, K., Kreinovich, V.: A linear-time algorithm that locates local extrema of a function of one variable from interval measurement results. Interval Comput. **4**, 176–194 (1993)
13. Zrilic, D.G.: Circuits and Systems Based on Delta Modulation: Linear Nonlinear and Mixed Mode Processing. Springer, Berlin, Heidelberg (2010)

Printed in the United States
By Bookmasters